Organize Your Stuff
Sustainable Ways to Declutter Your Space

Laura Jevtich

Crack of Noon Club

Copyright © 2023 Version 1

Copyright © 2026 Revised by Laura Jevtich Version 2

No part of this book may be reproduced in any form or by any electronic or mechanical means, including information storage and retrieval systems, without written permission from the author, except for the use of brief quotations in a book review.

This book is for informational purposes only. It is not intended to promote any violation of the law, and is true and complete to the best of the author's knowledge. The author disclaims any liability arising from the use of this information.

Cover Art by 100Covers

Editor Francesca Varela

LauraJevtich.com

Sign up for Laura Jevtich's Newsletter

All rights reserved.

Early Testimonials

By Virtue Meadow, "Wait—you made a casita in the desert with a 3,000 sq ft garage and a bidet? That's a whole reality show waiting to happen. Give us more!"

By Ester Williams, "The Gold Nuggets (aka What Made Me Actually Text My Sister About). That NPS Uniform Saga! The leather belt struggle? Perfection. I could feel you white-knuckling it before letting go." "Only the socks survived," should be embroidered on a damn pillow. That's your book's soul right there.

Plus, "It's like you're sitting across from me at a coffee shop, handing out legit advice about making cash from my clutter. That $15K house purge story? The stinky sneakers sale? Gold. I immediately trust you because you've clearly done this yourself. What's Killer: That opening line? Chef's kiss. You had me at, "money chapter." Your war stories (like the bidding war over used sneakers) are what make this actually useful instead of just another boring how-to. The photo tips? Pricing advice? So damn practical I wanted to start listing my crap on eBay immediately. You keep it real with lines like, "Your kids won't want your old stuff." Preach!"

By Stephanie Shackelford, "Well done! This is a good book, with a useful message and clear, concise information. This was very organized. It flowed well. It was written in a way that made it easy to follow, easy to remember,

ORGANIZE YOUR STUFF

and easy to implement. The repetition of principles is a great way to get a "lesson" across (says the teacher in me :)). It also would help to provide a mental framework that would help someone not be overwhelmed by the process. The examples and stories you wove throughout the book made it very personable. It was like listening to a friend give advice and stories. For many, those stories will help provide an "experience" that allows the reader to try something on his/her own."

By Brenda F, "Wow Laura, thank you so much for the organized way you did this! My little Type A self is nerding out!"

By Ash Reed, "you honest opinion here and throughout the book is GREAT! and much needed to about wasting time. WELL DONE!"

Dedication

The book *Organize Your Stuff* is dedicated to Carla Capuano, Tina Ventimiglia, Olga Jevtich, and Grandmothers on all sides of the family. They helped me in many ways, most likely without even realizing it.

And to Sasha, my husband, who created many of the mantras we live by while walking softly upon the earth.

Contents

1. Introduction — 1
   Four Principles: Keep, Recycle, Sell, and Donate or Dump

2. The First Principle — 5
   Keep

3. The Second Principle — 11
   Recycle

4. The Third Principle — 19
   Sell

5. The Fourth Principle — 25
   Donate or Dump

6. Medicines and Toiletries — 31
   Applying the Four Principles to your Bathroom

7. Clothing, Shoes, and Accessories — 47
   Applying the Four Principles to Your Wearables

8. Household Goods — 70
   Applying the Four Principles to Your Kitchen and Linens

9. Physical Media — 92
   Applying the Four Principles to Books, Maps, Photo Albums, Scrapbooks, Music, Videos, and Documents

10. Hobbies	117
Applying the Four Principles to Your Activities	
11. Tools for Home and Vehicles	130
Applying the Four Principles to Metal Doodads	
12. Electronics	142
Applying the Four Principles to Your Digital Life	
13. Food and Provisions	158
Applying the Four Principles to Your Supplies	
14. Ideas for the Future	174
15. Resources	181
Acknowledgements	183
About the author	185
Also by	188

Introduction

Four Principles: Keep, Recycle, Sell, and Donate or Dump

Whether you live in a house, an apartment, or an RV, downsizing to a smaller place can help you stay sane, save money, and enjoy life. In order to get started, you'll need to go through your stuff and get organized. In this book, I'll be helping you do that by sharing the Four Principles. These will help you decide whether to:

- Keep

- Recycle

- Sell

- Donate or Dump

Each principle has multiple tips to help you declutter your life and get organized.

But before going into the principles, I want to give you my perspective on why organizing your things is essential to your well-being.

Having an organizing system in place means my mental energy no longer needs to go toward the organization of things. I do not become frantic when I need to find my keys or any other items I need, because I know exactly where they are. Putting items in their proper place saves me time and money—I never lose my possessions or need to buy duplicates, so more of my brainpower can go toward creating and enjoying a colorful life.

My husband Sasha and I have lived in three different homes over the years. The first one was a semi-custom, 3,000-square-foot, three-story house with a two-car garage. I call this the big house. We went from the big house to a 24-square-foot RV in a matter of months. I call this the tiny home. While we made the tiny home work for almost three years, we eventually moved into a larger toy hauler RV. But I still call that one a tiny home, too, at 240 square feet.

During our thirteen years of living in tiny homes, we developed certain mantras, sayings, and ideas about living small or simple living, largely inspired by the book, *Your Money or Your Life: Transforming Your Relationship with Money and Achieving Financial Independence* by Joe Dominguez and Vicki Robin.

This book helped us realize that our possessions were ruling our lives, rather than the other way around. It also helped us realize that we had "life energy," a concept where you decide; do I pay X amount for this object, then how many hours of work, or life energy do I need to pay for it? If you decide there are too many hours to pay for the item, then why buy?

And so I got started, developing lists and maps showing the location of everything within the home. I measured the cabinets, shelves, and hanging places to fit two people into a 24-foot truck camper. Our covered utility trailer held motorcycles, extra water, seasonal clothing, and what I called "garage stuff."

We stopped the RV lifestyle in 2018, and I designed our current home, which I call the small casita, big garage. We live in the Mojave Desert and have an outdoor kitchen. With no oven inside the home, I designed the steel building to take advantage of the 2.2-acre property. Solar panels out front, the 3,000-square-foot building has two main parts: a 2,100-square-foot garage and a 900-square-foot casita home with one-foot insulated walls. The casita is our one-bath, one-bedroom home with a bidet (integrated toilet). We made this building (the garage and casita) our organized and easy to clean age-in-place home.

After years of learning how to become organized in my own life, I'm here to teach you how to do the same using the principles in the book.

Each chapter includes a section to organize your stuff. The smaller the square footage of a home, the more the need to organize.

One time a friend walked into my home and said, "I wish I could have my house looking this clean and organized. My home feels cluttered, and I'm frustrated by my family's inability to declutter. What do you suggest I do?"

Besides telling her to read this book, I suggested she change the way she communicates with her family. The best way to change a person's perspective is to help yourself first, then show them the achievements you've made. I also told her she should remove emotion from her discussions and decisions. Arguments occur when emotions are involved, and hurt feelings emerge. Nagging does not bring change—observance and a willingness to be respected does.

I also suggested that she start small and reward herself for each category or section completed by doing something fun with her family. This might mean playing in the yard or park during the summer, or taking a brisk walk around your neighborhood in the winter. Something that gives your family its undivided attention and gets everyone out of the house.

Then my friend asked, "Will this book help me if I am disorganized? What if I read the chapters backwards to forwards or jump around?" Not everyone reads in a linear fashion—I get that. At a minimum, I suggest reading about the Four Principles (Chapters Two through Five) and then reading the last chapter, "Ideas for the Future."

Other than that, feel free to skip to the types of items you need to organize first. The "Organize Your Stuff" chapters go from simple to complex in my mind, but that might be different for you.

The First Principle

Keep

The first time I experienced the downsizing process was when my husband Sasha and I moved from a 3,000-square-foot house into a 24-square-foot RV. It was a challenging process. One of our guiding principles was to place more value on experiences rather than on stuff, and to take our emotions and memories out of the equation. We had help doing this: my parents took or bought a few of our possessions, stating that we would get them back when we stopped RVing. We eventually got them back thirteen years later.

My husband and I are efficient together, so we have kept our sanity, time, and savings intact during the downsizing process, but some may not be so fortunate.

Many individuals retain their possessions for years. You may keep your things for decades if you buy objects that are durable, long-lasting, and have a long shelf life, especially if you take good care of them. Some of my sweaters and T-shirts have lasted over three decades. One of my laptops was over seven years old before I sold it.

It is commendable and environmentally conscious to buy durable and well-made items because you can keep them for a long time. However, avoid the trap of keeping for keeping's sake. Too many items mean you must keep, take care of, and store them.

Finding Your Why
Why does this book have relevance today? United States citizens migrate from town to city to suburb. When the economy encounters a recession, moving decreases. When the economy experiences a boom, moving increases. As the country's population becomes older and retires, downsizing and decluttering are inevitable.

Your family might be in the military, and they move from base to base all over the country. More often, I hear of adult children moving in with their parents, with either group taking care of the other. Making space for them means downsizing your possessions. Ultimately, you prepare for what you leave behind for your loved ones—and you must realize that they will not take much.

My Why
When I was growing up, my parents, aunt, uncle, and grandparents bought a summer home on the New Jersey shore. Three generations in a small five-bedroom, two-and-a-half-bath house; we made it work because we spent most of our time outside. During those summers together, my sister, cousin, and I would work at estate sales with our parents.

Their small business, called TLC, would do the complete package: clean, organize, and price the items for sale. They found places to donate the excess. TLC's estate sales were popular and spanned three days.

From the cumulation of years of experience with these childhood estate sales, the downsizing of our own home, and designing a sustainable tiny home, this book came to be designed, researched, and written. By following these steps, you too can feel sane, save some money, and get the time to enjoy your life.

Tip A: Similar Together

You decide what works best for you based on your storage and keeping your similar stuff together. By keeping similar items together, you can find everything in a logical place.

For example:

Clothing + Accessories = Bedroom

Toiletries + Medicine = Bathroom

Silverware + Plates = Kitchen or Dining Area

Food + Provisions = Kitchen, Garage, or Basement

Tools + Supplies = Garage

Hobby Items + Physical Media = In their unique places

In our big house, books were in every room. By going through these principles, I moved them into a central location. In our RV, we displayed them in one place—in the bedroom. I would switch out paperbacks when we went to a campground. Each new campground, we read new books.

In our small casita, I have a floor-to-ceiling bookcase. Every item in the bookcase has gone through the four principles system. Each item holds special meaning, does something useful, or is pretty. It is a colorful jumble of books, memorabilia, planners, and stationery.

Let's say you decide not to put your linens in the linen closet, as you don't have one. Or you use the linen closet as a pantry for your extra food or provisions (provisions are items bought but not used in the month bought), and instead, you place your winter linens in a box under the bed and put the summer linens in your closet. You design a new place for your items, which becomes the logical place.

Example

We put extra blankets, clothing, and linens in drawers built into the bed frame in our big house. Living in the RV, we had a tall closet between the bedroom and bath where we stored clothing and towels.

We live in a tiny casita with a large garage, with one full bath. The bathroom's lack of a closet door preserves the home's spacious feel. We store winter linens on the upper shelf of the closet in moth-proof bags, and we place other bedroom items in our bed frame drawers.

You might find unique places for storing your stuff. Look around and be innovative. What are rational places to store items? Would they be accessible? Have you got a ladder for reaching higher ground? Do you feel comfortable climbing a ladder? Can you crouch down to reach the lower drawer or back of the pantry? These questions will help you choose appropriate storage locations.

Tip B: Usefulness

I get it. All of your stuff is useful to you. Why have them unless they are useful? These questions, or a variation of them, appear within each chapter:

Have you or your family used it within the last month?

↪ If not, have you used it within the past six months?

↪ If not, have you used it within the past year?

↪ If not, why do you have it? Go to another principle.

A good rule to live by is that if you haven't used something in over a year, then the item goes. Some things, however, fall outside the one-year rule, such as belongings that have special significance. There are conditions for each of them within each subsequent chapter.

If you no longer want the item, you go to another principle. You can recycle, sell, donate, or dump.

Example: Winter Colds
One winter, my husband and I got horrible colds. In the large closet between the bedroom and bath of our toyhauler RV, I stored cough medicine, cough drops, and tissues. Some of the cough containers were half-filled with precious cough medicine. We were way out in the deserts of Arizona, over a hundred miles from the nearest store. Having those supplies made all the difference in our travels. They were useful to us. We kept them stored, and I knew where to find them.

Tip C: Storage and Containers

Once you decide to keep the stuff you own, make sure each item has the proper containers and storage bins. The items you keep will stay longer, hold their shape, and not get torn or stained.

Examples
We usually buy food in bulk, so I keep it in food-safe storage containers. And to minimize closet clutter, I utilize clear, heavy-duty storage containers to switch out my winter and summer clothes. In the desert, summer clothing means April through November.

Different containers and storage bins will be used in each category. One essential nugget to remember is that you should always make sure you have enough room for all your stuff, with some air between the items or containers.

If you lack the space, realize that letting go of your belongings will mean Recycle (Second Principle), Sell (Third Principle), or Donate or Dump (Fourth Principle).

At the end of each chapter, after the principles, you will find examples of how to organize each category, plus worksheets to write where everything is located.

The Second Principle
Recycle

Recycling your stuff can help you save money, and it can help you make money, too! Recycling is not about being Captain Planet—it's about being smart with your stuff and using things in new and innovative ways.

For example, one creative way to recycle your stuff is by upcycling. Upcycling is when you turn an item you do not need anymore into a product of higher quality or value than the original. Several of my artist friends upcycle their bits and pieces to make jewelry.

This chapter and the subsequent chapters explain other ways to recycle your stuff—ways that help keep you sane, create more time, and save more money.

There are seven R's of Recycling for Sustainability. Each chapter has at least three recommendations and suggestions.

The Seven R's of Recycling for Ecological Sustainability

1. Recycle

2. Refresh

3. Repurpose

4. Reuse

5. Reduce

6. Repair

7. Refuse, Reject, Rethink, and Regift

What is sustainability? Essentially, sustainability means protecting our natural resources in order to ensure ecological health and stability for generations to come.

This equates to living small and softly upon the Earth. By knowing what is below our feet and in our air, we can make sure never to destroy either. Working as an environmental geologist who cleaned groundwater, soil, and air, I know firsthand how difficult the job can be.

Both words: sustainability and recycling—have come under attack in recent years. New research and books, such as *The Carbon Almanac*, give examples of sustainability in all aspects of life. They define sustainability as "producing in a way that does not use what can't be replaced and does not damage the environment" and recycling as "processing something, usually waste, to use again, in the same form or by creating new material."

Consider recycling and sustainability this way: we recycle an object until it becomes nothing or until it's reverted back to its raw material. We recycle in order to avoid the repercussions of a destroyed earth, loss of natural resources, and to make us feel as though we are helping the environment. Sustainability is the overall goal of having resources available for future generations and the ability to care for and nurture nature and the environment—ideas which are the credos of the National Park Service, the Bureau of Land Management, and the U.S. Forest Service.

Tip A: Recycle

Recycling means converting waste into reusable materials that can be used again. Most people think of the word recycle as those bins for newspapers, paper, aluminum cans, and sometimes for magazines, glass, and plastic. However, there are a lot of other things that can be recycled in your home before throwing the item out.

Examples

- Recycle your plastic grocery bags by taking them back to the store, or repurpose them as trash bags, reusing them again—all for free.
- Wash a glass tomato sauce jar and use it for sprouting seeds, creating to-go salads, containers for small objects, and as glassware.

Example: Upcycling

I went to a wedding with a farmhouse-chic theme, and they used glass jars as tea candle holders. Wedding magazines picked up her Pinterest wedding photos for years because they showcased inexpensive, chic decorating ideas. She gave those glass jars to friends so they too could have Pinterest-worthy weddings.

Caution!

The trouble with recycling is that every community has different programs. Certain plastics, especially the high-numbered ones, need robust recycling centers. Glass needs high temperatures to melt and reshape. Most communities do not recycle glass anymore because of contamination from other recycled items. Glass is also heavy and breakable, leading to higher costs in transportation. And finally, recycled glass does not have the market to recycle, making it less economically viable.

Tip B: Refresh

Refresh isn't a fancy concept. It's sharpening your knives so they last longer than most relationships. Refresh means to reinvigorate or give new strength to an item. Many people refresh their clothing by dyeing a shirt or sweater. They remove fabric pilling from a sweater. They put new soles on their Crocs to make them last longer. One of the best uses for small pins or bows is to place them on your shoes to give them a new look with flair and shine.

Examples

- Use a permanent marker to remove scuffs on shoes or purses—this costs much less than buying new.

- Put a little WD-40 on a tool to make it work smoother.

- Rebuild a computer or laptop to make it faster and help it store more of your important documents and pictures.

The ideas are endless and touched on throughout the book.

Tip C: Repurpose

Repurpose means to use an item that you would otherwise get rid of in a new way, making it useful again. This is one of the best R's of Recycling for Sustainability, but it ranks among the most arduous tasks.

Examples
Clothing is an obvious choice:

- Pants convert into skirts.

- Shirts convert into scarves, quilts, or soft rags.

- Frayed blankets convert to bath towels, bath towels convert into dish towels, and dish towels transform into wash rags.

I read a Yiddish folk story and song called "I Had A Little Overcoat" about how a young man and his mother were resourceful. The story goes on that his mother made different unique items for him out of one overcoat, turning it into a blanket and eventually a button. When the young man loses the button, he makes a song. This story highlights resourcefulness.

And it's not just clothes—we can also repurpose food into several possibilities. A roasted chicken can become chicken soup or chicken salad. While I have no cookbook in this book, if you start with a particular food item, you can repurpose the leftovers into another simple meal.

There are many ideas for each category, especially if you have a house and can store stuff. You might have to increase your recycling skills if there is limited space.

Tip D: Reuse

We constantly reuse our glassware and plates without thought. Reusing means getting the most out of your items before you recycle, reduce, repurpose, or refresh.

Reuse can also mean buying something secondhand. Thrifting items allow you to purchase secondhand or even brand-new items at lower prices while you give money to worthwhile organizations.

Use your products every day. It could be something as simple as using cloth shopping bags at the grocery store, or something more complex like buying in food bulk and reusing your containers for food storage.

Examples

- Visit the library to borrow books, music, videos, magazines, and newspapers

- Buy most of your clothes from thrift stores.

- Water bottles, stainless steel, or BPA-free plastic are much better than buying plastic water containers.

- Buy food with less packaging, called bulk buying, where you use the bags provided to bring the items home and reuse your containers.

Many reuse possibilities exist; the list is almost endless. You will see examples throughout the book.

Tip E: Reduce

Reduce the amount of stuff you have, or will have in the future, by owning less. This saves the environment, your money, time, and sanity. Buy durable products that can withstand pressure, wear, and damage.

Examples

- Don't buy paper goods; instead, have durable plates and cloth napkins.

- Buy in bulk.

- Borrow infrequent-used items from friends and family.

- Compost your yard waste.

Using cloth napkins makes every meal feel classy, and they are negligible in the wash.

Example: Upcycling

We started our RV adventures using a gift. Most of our family and friends in the Washington, D.C. area did not understand RVing. They thought we were camping and picnicking. One friend of that mindset gave us a picnic backpack filled with reusable thick plasticware, Corelle or enamelware plates, cloth napkins, a matching tablecloth, two thick plastic wine glasses, and a wine opener. These items became the basis of our kitchen during our first trips in the truck camper. We used the backpack to cart groceries to the RV.

Tip F: Repair

While my sewing is not up to par with the older generations, I can still thread a needle and close a tear in my pants. Buttons are harder but still doable, and they keep my shirts from popping open. To repair is to restore something damaged, faulty, or worn to an acceptable condition.

Examples

- Fix a broken chair with strong glue.

- Clean an item to make it work again, such as computer mice and keyboards.

- Sew or bring an item to a seamstress or tailor to adjust your clothing.

Tip G: Refuse, Reject, Rethink, Regift

Refuse, reject, and rethink tips are your mind-over-matter ideals. It is your decision. You are in control of what you buy and store in your home. This includes gifted items, whether free or given to you. Just because they gave it does not mean you have to keep it. The Seinfeld episode "The Label Maker" is a classic example of **regifting**.

You **refuse** to buy disposable plastic items. Items like plastic bags, straws, bottles, and utensils will no longer be in your home.

You **reject** buying things you do not need. If your neighbor has ten matching plates and glasses, you borrow them. You can then clean and return them after you are done.

Rethinking is how you look at your possessions, manage, organize, and decide to dispose of your stuff.

The Earth doesn't need perfect recyclers; it needs people who try. So the next time you're about to throw something in the trash, ask yourself: "Which 'R' can I use first?" Then go do it.

Chapter 14: Items and Ideas for the Future has a discussion on how you can rethink your personal relationship with your stuff.

The Third Principle

Sell

This is the money chapter. Got stuff that's too nice to dump but just collecting dust? Time to turn your clutter into cash. Maybe you received duplicates, or your housemates have similar items. This chapter is where you learn how to sell them.

As Sasha says: "Keep the best and get rid of the rest!"

Example

Before putting our big house up for sale, I sold and donated over $15,000 worth of items we no longer needed.

When we lived in our RV, my husband and I continued to organize, sell, and donate items. Our motto was: "If you bring something in, you must take something out."

Being in our sustainable casita home now, we tend not to buy or bring in extraneous items. I would rather not clean or store items, especially if they are decorative. My crafts and planners get the decorations; the interior of my home does not.

Tip A: Take Pictures of Your Stuff

Yes, a picture is worth a thousand words. By taking distinct pictures, your items could sell for more money than you thought. How you take a picture is essential.

If people are looking for a particular color of an object, adding filters harms your price and your honest online profile. Not everyone's screen displays the same colors. However, allowing potential customers to see your goods can help you make more money. Here are some general tips:

- Try to stage your stuff on a white background.

- Photograph only the item being sold. Otherwise, people get disappointed and distracted.

- A small object needs four pictures: one of the front, one of the back, at least one of the sides, and one of the bottom.

- Medium items, such as computers, should have a minimum of six pictures.

- Large items, such as furniture or cars, should have at least ten pictures.

The mantra, *better photos lead to better prices* is possible with correct pricing. I cannot always guarantee it, but the saying has worked for us.

Example

I sold a pair of sneakers on eBay back in 2003 or 2004, the heyday of selling anything on eBay. We took the maximum number of pictures and wrote a blurb worthy of several laughs. We titled the sale, Worn But Not Stinky Sneakers. At the end of the auction, there was a bidding war, and the sneakers sold for $25. Since that time, anything we sell on any platform gets the maximum number of photographs.

Tip B: Know the Worth of Your Stuff

When you sell, you need to know the original price and worth of your stuff in the secondary market. Gone are the days of one item with sparse descriptions and one photograph.

Example

Remember those sneakers? Here's why they sold for $25 and why that wouldn't happen today. The pictures were spot on, of course, but the real reason? I researched eBay looking for size 8.5 sneakers. There were hardly any available.

Rule #1 on eBay in 2006: sell what people want. Rule #2: start the bidding at $1. Yes, you read that correctly. Everything we sold on eBay during that time started with a $1 price point. People had to pay for shipping, and we gave them the lowest price point for shipping. It made our listing almost too good to be true.

And why wouldn't this happen today? Of the fourteen thousand listings for size 8.5 sneakers, only fifty were auctions. People are selling items on eBay as a business. We weren't. We only wanted to get rid of things we couldn't use anymore.

Look on eBay, Poshmark, and Mercari to find your items in the sold listings. Watch to see if your goods are selling. Some things sell better at certain times of the year and in certain parts of the country.

- Do not sell winter clothing in March or April. You will not get much money for them. (Unless you live in the southern hemisphere—then the opposite holds.)

- Sell children's items in late July and August before school starts.

- Only try to sell camping gear in late fall if you live in the southwestern states such as Arizona, south Texas, southern Nevada, or southern California. Otherwise, wait until spring to sell outdoorsy items, unless they are hunting items.

- The best time to sell a house is between May 1st and May 15th. You can get almost one percent higher price during those two weeks. Why? The reason? Those dates are before most schools' end.

Price your goods based on the adage: most clothing, electronics, furniture, and housewares will sell for one-tenth the retail or sales price you bought. Why is this the case? Most of these types of items have a seasonal or half-year shelf life. The companies bring out new and improved things throughout the year, so your selling prices must be lower.

The adage, though not absolute, guides pricing decisions.

Do not put emotion into selling your stuff.

Do not believe the hype! Your children or other family members will not want your old things.

Example

When selling electronic goods, my husband and I have found Apple products sell for the best prices. Your iPhone (pick a number) will sell best when the phone is two years old. Once the phone gets to the four-year mark, selling it back to the company will get you the highest amount.

Tip C: Where to Sell Your Stuff

Most people think of garage sales to sell their items. Here is a list of familiar places to sell:

- eBay: sell small things that will not incur significant shipping costs and make sure you are okay with the massive fees to sell items.

- Poshmark and Mercari are other places to sell clothing, crafts, or hobby supplies.

- Craigslist is for furniture and vehicles. You will not get much for your items, but they can go faster if you price them right and have pictures.

- Amazon is for books, music, videos, electronics, and anything with an SKU number. An SKU number is the barcode on your item. Look on Amazon and then sell your stuff. You want to make sure your cost to ship is inexpensive. Otherwise, you will lose money on the sale.

- Estate sales are a great place to sell antiques, household goods, tools, and artwork. Estate sales work the best where antiques sell well, such as New York, New Jersey, District of Columbia, Virginia, Maryland, the New England states, and some places in Texas, Chicago, and San Francisco. Antiques command top prices in established markets with classic, larger homes.

- Facebook groups and Facebook Marketplace are good places to sell things like crafts, tools, and other inexpensive goods. Buy/sell/trade groups in your local area can number in the hundreds.

- Consignment stores might be among the best places to sell your stuff, if you have the time.

- RVTrader.com is a good place to sell your RV and RV-related items. These are by consignment.

- If you live in the Southwestern U.S., you can use KSL.com to sell cars, RVs, and other equipment.

Example

My husband and I have sold over five vehicles over the past twenty years, as well as one RV and countless other items using a variety of methods. I sold hundreds of items on Mercari back when they had low fees. Craigslist, along with Facebook Marketplace, helped us sell our fifth-wheel toyhauler for the same price we bought it. The same was true of our F450 Ford hauler truck. Recently, I helped sell antique costume jewelry at a community garage sale, where we made over $300.

Find places where you can sell. Start small, with a friend or family member. Know your area and town. Do they allow garage sales every weekend, or only once a year? Don't worry, you got this, and each chapter will help you make some money on your stuff.

The Fourth Principle

Donate or Dump

When all else fails, donate. I am not sure if this was our mantra or something we picked up while emptying our big house. However, we kept using the sentiment throughout all our moves and homes. Donating means giving unneeded items to charity shops or thrift stores, or passing things on to friends and family members who need them, such as donating old furniture to your child as they move into their first apartment.

Donate: When All Else Fails

You went through the first three principles: keep, recycle, and sell. None of them worked for your item. The next best thing you can do is donate.

Example

I donate to our local library, museums, community parks, and other local businesses. When we sold our big house, we donated posters and maps to several national parks. The newly updated offices needed wall art. We gave them several framed posters and a unique map of the Potomac River created by a local artist. Thrilled to receive such a map, the Chief of Interpretation put it in her office.

Some other ways to donate:

- Your local thrift store is the best place to donate before you go to the big names. You'll begin a relationship with the store. It may be an organization you feel is worthwhile and needs your stuff.

- The Salvation Army is a nonprofit and one of the largest thrift store chains in the United States.

- Goodwill is also a nonprofit. Their mission is to give jobs to those with disabilities, senior citizens, second-chance individuals, veterans, and youths.

- Donate to libraries, national parks, hospitals, food or soup kitchens, or your child's school.

- Animal shelters regularly have donation and sale days. Ask your veterinarian about the best ones in your area.

- The BuyNothingProject.org site focuses on donating items within the community. They are on Facebook, and they have an app.

Other websites include Freecycle, Nextdoor, United Way, Red Cross, and hunger relief programs.

Make sure to keep excellent records, especially if you donate over sixty percent of your adjusted gross income. You should protect yourself against having to pay taxes. Your best bet is to check with the IRS.gov to protect yourself. Read the IRS rules before tax season kicks you to the curb.

Donating isn't about guilt; it's about your stuff becoming someone else's treasures. Donate the appropriate items to their rightful places. More information about where to donate is available throughout the book.

Dump

When you can't donate, your last option is to dump. Dump means to discard something useless or unwanted. When you throw an item away, it goes from your trash can to the landfill, where it gets incinerated or is used to generate energy. Garbage used to be separated into wet and dry waste. Now anything you dump equals municipal solid waste, which sounds disgusting, and most people will strike it from their brains as it sounds too much like something else.

Imagine for a moment your dump items are no longer picked up by a service. Instead, you bring them to a landfill. They charge by the pound. How will you reduce the landfill volume?

- Read over the specific chapters following this one.

- Make sure nothing valuable is in your trash.

Some neighborhoods have a "junk day" where you can put your stuff at the curb. In the evenings, people go around and take items they can fix up and sell. It is a beautiful way to get others to repurpose and recycle rather than dump.

- See if you can put a box labeled "FREE" near the street. It might be gone the next day.

- Before discarding items, inspect thoroughly for anything valuable.

Tip A: How Long to Decompose?

It is impressive when you think of how long it takes something to decompose. Decomposing means the item breaks down, becomes rotten, or moves toward simple elements. Each category will show how long certain things take to break down.

Tip B: Resources and Links

Supplementary materials are located here. Books and or websites that can give you more help or insight into why I suggest not dumping an item.

Summary of the Four Principles

There you have it, the Four Principles of organizing your stuff to keep you sane, give back your time, and save money. There are chapters on the individual categories of your stuff, starting with medicine and toiletries. Prior to delving into the first category, here are some handy hints to kick-start your journey.

When starting this process, try to do it when you are at your best. Start early when you are refreshed and ready for the chore. Have three laundry baskets and a trash box. Label or color-code each basket for the three principles of keep, recycle, and sell. No label needed for the donate or dump box.

Have your cell phone or camera ready for the pictures. Take the time to remember your why and let yourself imagine how good it will feel when you get back your time, save yourself some money, and stay sane during the process.

WORD of CAUTION

Any time during this process, with any of the chapters, if you feel yourself getting overwhelmed, stop what you are doing, sit down, and take some deep breaths. This happens when overwhelm sets in, and you might destroy objects or say something to your loved ones that you will regret later.

If you get overwhelmed each time you try to start this project, then start small. Forget putting everything in a centralized place. Instead, get your laundry baskets and dump box and go through the process with just one drawer.

Once the feelings of overwhelm and defeat pass, give yourself a mental shake by putting on music, having a refreshing drink, or getting some air. Take frequent breaks throughout the day.

Or use the Pomodoro Technique:

1. Using a kitchen timer or your smartphone, set the timer for twenty-five minutes.

2. Work on organizing one drawer.

3. End work when the timer rings.

4. Take a quick break for five minutes.

Do this technique four times if you are able.

1. After four Pomodoros, take a longer break of thirty minutes.

2. Once the long break finishes, return to step one.

If you need more words of encouragement, email me at <u>Laura@Lauraevtich.com</u> and I will send you pictures of organized shelves in medicine cabinets or tips on how we store extra shampoos and soaps. If you want to see my bookshelf, which is NOT organized, watch a video or two of mine on YouTube.com/@LauraJevtich.

You can do this!

Medicines and Toiletries

Applying the Four Principles to your Bathroom

Organizing medicines and toiletries usually starts in the linen closet and bathroom area. Many people have so much stuff in their drawers, cabinets, and closets that they cannot close them. Together, we will go through the four principles in this chapter so you can finally have an organization system that keeps you sane, saves you money, and gives you more time to enjoy your life.

Before you begin, I suggest putting everything in one centralized place for each category and item. Then you can go through the steps.

- Place everything on a table or bed in a centralized location.
- Have adequate lighting to see everything.
- Get comfortable with it all.
- If the medicine and toiletries spill over to the chairs or floor, try doing those items first.

This reveals your extraneous stuff. Laying out all of your items together in one place allows you to see everything you have, including duplicates, empties, or trash.

The First Principle: Keep

Typically, most people keep most of their medicines and toiletries. When your stuff starts to overwhelm and overflow, many people go a little crazy. The best way to stay organized is to have everything in its spot.

Tip A: Similar Together

Put all like items together on the table or floor to view everything in this category.

- Makeup
- Hygienic products (shampoo, conditioners, soap, deodorant, etc.)
- First-aid supplies (hydrogen peroxide, bandages, creams, etc.)
- Dental products (toothpaste, toothbrushes, floss, etc.)
- Medicines (cold or flu medication, pain medication, and prescriptions)
- Bathroom paper products (tissues, toilet paper, baby wipes, etc.)
- Feminine products

Each item within the category gets a separate area at your table, and the next step will be to combine all like items. Make sure you screw the tops on correctly to prevent spills.

First, verify that no makeup, chapstick, or medicine has expired. Some medications, makeup, and vitamins have expiration dates. Does it have the same consistency, look, color, smell, and texture as it did when it was a new product?

Example

I had organic chapsticks, although I did not realize they were organic. After several weeks of using them on and off, I noticed some black around the rim. I researched why they had black edges, and I eventually decided they were no longer in the keep section! I scooped out the goo and recycled the plastic.

If you are uncertain about a particular drug, look it up on WebMD.com or MayoClinic.org. If you are still concerned, visit the nearest pharmacy.

Check each item to ensure there are no punctures or openings in the medicines and toiletries. Keeping the items closed will guarantee longer freshness.

Separate any expired items by placing them in a unique area on your sorting table. Proceed to Principle Four in this chapter for those items.

Tip B: Usefulness

Go through your medicines and toiletries, using this list for each one or each set:

Have you used this item in the past month?

↪ If not, have you used it in the past six months?

↪ If not, have you used it in the past year?

↪ If not, have you kept it on hand for any other reason?

↪If not, why do you have it? Put it into the Second Principle laundry basket if you no longer use it.

Example

We use hydrogen peroxide and Dr. Bronner's Castile Liquid Soap daily. These products have diverse applications, as illustrated below.

Half water, half hydrogen peroxide:

- Use as a homemade mouthwash for teeth and gums.

- Use as a cleaning solution for toothbrushes, sponges, and countertops.

- Add to bathwater to soak your feet and soften your calluses.

Full-strength hydrogen peroxide:

- Use ten drops in an ear to remove earwax.

- Clean your Neti Pot and other surfaces when water is at a premium.

- Put on a cotton ball or gauze to clean wounds.

- Great for cleaning everything from tile to toilets; let it sit for thirty minutes. Test a small area first before using it.

- At less than $1 a quart, it is a cheaper and easier storage alternative than most cleaning products.

Example: Dr. Bronner's Pure Castile Liquid Soap

I started buying Dr. Bronner's Liquid Soap in 2010 when I worked at Canyonlands National Park. The natural food store had it in bulk, or you

MEDICINES AND TOILETRIES

could buy it by the quart. I purchased my first quart of peppermint soap back then and have repurchased a quart every year since.

We use it as a shampoo, soap, and for cleaning. It is biodegradable, and the labeling of the bottle is quite interesting. You dilute it and use it for so many purposes. The quart usually lasts us at least a year, and now we buy from their website.

Dr. Bronner's Pure Castile Liquid Soap label:

> Dilute: Enjoy one soap for 18 different uses! Shave-Shampoo-Shower-Bath-Mop-Launder-De-grease! What are the "18 in 1" uses? You can use Dr. Bronner's soaps for washing your face, body, hands, and hair, for bathing, shaving, brushing your teeth, rinsing fruit, aromatherapy, washing dishes by hand, doing laundry, mopping floors, all-purpose cleaning, washing windows, scrubbing toilets, washing dogs, controlling dust mites, and killing ants and aphids.

Tip C: Storage

Using the drawers or cabinets you already have will help you determine what you need in storage containers. When you measure the most significant storage areas, you can figure out where to store your items. Tray inserts work very well on shelves and drawers. You can buy closed, see-through, stackable containers, and you can use them to hold pill containers or other loose items in one place.

Other containers include:

- White, multipurpose plastic storage baskets
- Clear storage or latch boxes in small, medium, and large sizes.

Our medicines and toiletries are in two places in all the homes: in the kitchen and in the bathroom. I keep only a limited supply of toilet paper on hand for guests who do not want to use the bidet, as they are bulky items taking up space. They are also cheaper than other medicines and toiletries, and I can store them in the garage or containers.

The Second Principle: Recycle

You might think you cannot recycle your medicines or toiletries, but this is untrue! I recycle most of my containers, including Dr. Bronner's soap bottles, prescription bottles, and other containers.

Tip A: Refresh

Store items in small, reusable containers that are easy to reach and access. Keep bulk supplies of these items in larger containers stored out of sight, such as in a closet or under the sink. Then, you can restore your collection once or twice a month as needed.

Tip B: Repurpose

Check Google or frugal paperback books to find other ways to repurpose items.

- Baby wipes can be used for cleaning your body when water is scarce.

- When we lived in tiny homes, water was at a premium. Dr. Brommer's '18-in-1' Hemp Peppermint Pure Liquid Soap has sixteen uses besides shampoo and soap. Same idea with hydrogen peroxide, baking soda, and nail polish remover.

MEDICINES AND TOILETRIES

- A pair of sharp haircutting scissors can cut through more than hair—they also cut through gauze, bandages, cloth, and some plastic.

Tip C: Reuse

Most medicines and toiletries get used up, eventually. However, we can continue to reuse the containers and packaging they came in and buy replacements in bulk.

Examples

- Some people reuse toilet paper tubes and tissue boxes as starters for kindling.

- You can reuse medicine containers for other medicines, but mark them and wash them out first. Or you can use those containers for items such as nails, screws, buttons, and other loose objects.

- My prescription containers become dispensers for crystallized ginger, which I use to settle my stomach. Those containers are the perfect size and do not pop open.

- You can find new uses for toiletries. For example, body wash can be used as shampoo as long as you do not have sensitive skin.

The Third Principle: Sell

Other items to sell include unopened containers of toothpaste, toothbrushes, hair products, almost anything in a container. You can sell makeup and personal items as long as they are NOT open and have not expired. Some of the best places to sell will be your garage, yard, or estate sales. As soon as you try to sell at vendor places where you need to pay, you will run into problems with opened containers. Best to not sell them there. You still need to go through the tips below: a) take pictures, b) find the worth, and then c) sell.

Tip A: Take Pictures

You do not need to take pictures when selling at garage, estate, or flea markets. Before our big house sold, I held garage sales and sold my ample supply of makeup and toiletries.

If you sell online, you need to take pictures. Ensure the area where you take the photos is well-lit and contrasting (white containers on a dark background or dark containers on a white or off-white background). It should be clean and free of clutter. Taking a picture with multiple items confuses people.

Take at least four pictures of little items—top, bottom, and sides. Take pictures to note the label, price, an expiration date, or best-by dates. The pictures of the unopened packaging are key. Take at least six to eight pictures of oversized items, especially if some features need a close-up.

Tip B: Find the Worth

Look at local thrift or garage sales in your area or on Amazon, eBay, and discount stores. The worth of the products depends on several factors, such as how much is still usable, the original price, the age of the product, whether it's in a reusable container, and what the brand is.

Tip C: Where to Sell?

You can sell unopened items on eBay and Amazon. I know people sell Avon or Mary Kay samples and products at flea markets and garage sales. Realize that you may sell these at a lower price than bought. Makeup and other toiletries sell for pennies on the dollar. If the items are unopened, they usually sell for one-quarter of the original price or less.

Example
Avon's Skin So Soft sells well at garage sales or thrift stores. The owners are selling them at a loss. I bought Skin So Soft for $1 a container. Unless a representative got the item for free, selling the products at a garage sale may not make sense. Selling unopened containers on Amazon could get you much higher prices, or you could sell at flea markets and vendor sales.

The only way to make money on toiletries is bulk buying or getting items for free.

CAUTION!

A word of caution about flea markets: they hold higher prices than garage sales because a vendor has to pay for the spot. However, you might find a swap meet or flea market where you can sell your items for cheaper in the long run than online. If you want to sell items without major fees, your best bet is a garage sale. Check your local ordinances—some cities limit the number of sales per year or how you advertise the sale.

Revolving around individual factors, vendors and flea markets ask for a vendor's fee, costing anywhere from ten to hundreds of dollars. Most ask for an application and have a form for you to fill out for the state's sales tax. This means you must charge sales tax on your items and send your state a check for the sales tax. This is especially true of new things. For used items, I would suggest going to a swap meet rather than a vendor-type sale.

The Fourth Principle: Donate or Dump

Tip A: Donate

Many worthwhile shelters, military, and transitional places need toiletries for their clients. Find the best ones in your hometown. They usually need personal-sized shampoo, conditioner, soaps, tissues, deodorant, toothpaste, toothbrushes, razors, shaving cream, lotions, feminine hygiene products, and sewing kits.

Tip B: Where to Dump

There has been much in the news about people throwing away their medicines by flushing them down the toilet. This is a terrible idea. If you do not finish your medicine or find it no longer helpful, bring it back to your pharmacy or doctor. Some police stations also have places for medications.

The FDA.gov has the best recommendation:

- Medicine take-back programs by the Drug Enforcement Administration (DEA) tend to have national prescription drug take-back events. They have two each year in April and October. The link will tell you where and when you can take back your medicines or prescriptions. www.deadiversion.usdoj.gov/drug_disposal/takeback/index.html

- Bring the medicines to pharmacies, hospitals, or law enforcement agencies.

Example

If you cannot do either of the FDA recommendations, try this: mix your unneeded medicines without crushing them or mixing them with any non-medications. Put them into sealable plastic bags. Throw the bags into your garbage. Scratch off or mark off any personal information on your containers. You can then reuse or recycle these.

You may have to charge sales tax at flea markets and swap meets, but you might not need to charge sales tax when selling at your home. You want to ensure you are selling your personal, used items at a loss. Otherwise, you may have to pay sales tax or pay the IRS for the income received from your garage sale business.

Tip C: Resources

With regard to selling your toiletries to the public, more than likely this will count as a small business. If you sell from eBay, Amazon, Facebook Marketplace, or other peer-to-peer resale places, any sales over $600 mean an IRS form 1099.

- IRS: https://www.irs.gov/businesses/understanding-your-form-1099-k

- Nolo: https://www.nolo.com/legal-encyclopedia/taxes-when-you-sell-things-online.html

How To Organize Your Medicine and Toiletries

If you take medicine or supplements, get a weekly pill container and put in a month's worth. Put the container close to your eating area to remember to take your pills. Put the excess pill containers in a see-through box and place it in the kitchen cabinet.

For medicines taken infrequently, put them all into see-through, stackable covered boxes and place them in your bathroom shelves or drawers.

One-bathroom home: Storage may be tight with two people. Each person gets a drawer in the bathroom. They can store whatever they want. The rest of the bathroom needs organizing. Under the sink, place a small shelving unit to store loose items. Then, add the equipment needed to clean the bathroom and any cleaning supplies. Make sure nothing is near the sink pipes. Place soft items (tissues and toilet paper) away from the cleaning supplies.

Having shelving or a closet within the bathroom is ideal for storage and organizing. Each shelf could hold hygienic, dental hygiene, and feminine products, and first aid supplies. Using clear, stackable boxes or shoe boxes allows you to see what is in them and get to the items quicker. You decide which products take priority on which shelf to store them.

Two-or-more-bathroom home: Designate one bathroom for the cleaning supplies. Have first-aid kits and supplies in each bathroom. Each person who uses the bathroom gets a drawer for whatever they want. You can also decide which bathroom has the soft items, as well as storing other products.

Garage space: If your home has garage space available, build shelving going to the ceiling. Make sure the shelving is at least five inches deep, the size of a paper towel. If your laundry area is in the garage or a separate room, use this space for your extra supplies for medicines and toiletries.

The basement could be an excellent place to store scented or sweet items; have secure, airtight storage containers to deter pests. Extra hydrogen peroxide, unscented liquid soaps, and cleaning supplies work well for storage in the basement. If you have an attic, put only soft items in lockable tubs. You do NOT want rodents getting in. Or store in the basement and place it on the higher shelves.

MEDICINES AND TOILETRIES

How To Organize Your Meds & Toiletries
Here are some suggestions for the worksheets below.

Keep	Weekly medicines	Breakfast cabinet
Keep	Months worth toothpaste	Linen closet
Recycle	Hotel shampoos	Refresh
Sell	Gifted makeup	Poshmark / $15
Donate	Oral care	VFW
Dump	Expired meds	Doc or Pharmacy
Dump	Toothpaste tubes	Terracycle.com
Dump	Pill Bottles	Ridwell.com

You can find these tables by signing up for my email newsletter at LauraJevtich.com/OYSV2 and you can print these out over and over.

Organize Your Medicines & Toiletries				
Keep		Recycle		
Item	Location	Item	Location	Repurpose, Repair, Refresh

MEDICINES AND TOILETRIES

Organize Your Medicines & Toiletries			
Sell		Sell	
Item	Location	Item	Location

Organize Your Medicines & Toiletries				
Donate		Dump		
Item	Location	Item	Location	Repurpose, Repair, Refresh

Clothing, Shoes, and Accessories

Applying the Four Principles to Your Wearables

I used to own fourteen pairs of underwear so I would only need to wash my clothes twice a month. I also owned pantsuits in seven different colors, hoping they would help me feel confident. It took me a long time to find I didn't need all the colors of the rainbow in my closet, and my confidence soared. Here's how clothing stopped being my security blanket.

Starting the organizational process with something easy such as the medicines and toiletries gives you a *win* sooner. We can get into the meatier categories because you understand the process.

One of the best ideas is called a capsule wardrobe. This works well for office people and minimalists or simple lifestyle advocates. I will explain it further at the end of the chapter.

The First Principle: Keep

Many people keep their clothing, shoes, and accessories for long-term wear. I know some fashionistas who keep up with the latest trends. Few possess sufficient storage for their wardrobes. Some of us keep clothing for almost too long, not realizing we are never wearing certain items.

Example

I worked as a park ranger for the National Park Service (NPS) for five years. I amassed over seven full uniforms—two for winter and five for summer. Once I quit working for the parks, I had to decide what to do with the uniforms.

Throughout this chapter, these uniforms serve as examples. Incidentally, discarding those uniforms took five years of deliberation. Carrying something around that long drained my sanity, causing me to lose precious hours, and it cost us money by having extra weight in the rig.

Just like that pair of jeans you "might" wear, think of clothing, shoes, and accessories as Band-Aids that need to be ripped off quickly. It will be less painful in the end. And there will be fewer moths in your closets and drawers.

Tip A: Similar Together

Take out all the items. Start with your stuff first, then work with other members of your family. Put similar items together, such as:

- T-shirts and shorts

- Long-sleeve shirts

- Sweaters and sweatshirts

- Three-quarter sleeve shirts

- Hanging shirts, dress shirts, and dressy blouses

- Pants, slacks, jeans, and capri pants

CLOTHING, SHOES, AND ACCESSORIES

- Dresses and skirts
- Vests, jackets, and coats
- Business suits
- Undergarments and socks
- Bathing suits and cover-ups
- Sneakers and tennis shoes
- Hiking boots
- Slip-on shoes and dress shoes
- Jewelry
- Scarves, ties, bandanas, and hats

You might also decide to separate certain items as expensive or inexpensive. Put similar items together on your table or bed. Figure out all of your available storage places—drawers, cabinets, closets, and containers. If you realize you do not have the space, you have two options:

- Look for storage in unlikely areas.
- Look to the Second or Third Principle.

Try to put your stuff in the most logical space. If you have a closet, put your hanging clothes in there. If you have clothing you wear only once or twice a year, put it in moth-proof bags or boxes and place it under your bed.

Example

When we started the RVing lifestyle, my off-season clothes went into see-through boxes. I would remove all the summer items, clean them, and then fold or assemble everything. The heaviest items went to the bottom, and the lightest things, like socks, were on the top. Then, once the season changed, I would get all the winter items cleaned, folded, and put into the boxes comparable to before, heaviest to lightest. Sealed boxes and cedar blocks kept moths out.

Tip B: Usefulness

How often do you wear your clothes? Going through a process works best.

Look at the first item and figure out when you wore it last:

Has it been worn in the past month?

↪ If not, then has it been worn in the past three months?

↪ If not, then has it been worn in the past six months?

↪ If not, then has it been worn in the past year?

↪ If not, then is it a unique item? Are there unnecessary duplicates? Do you only wear it at weddings or funerals? Is it clean, and are no repairs needed?

Put the thing aside for something unique. Review your wardrobe.

You have three baskets: keep, recycle, sell. The keep basket will fill up first. Try to go through everything in one section of the room, either your closet or a bureau. This way, you won't feel overwhelmed. The last container should be a box, as it is for donate or dump. Since you give it away for free, no sense in giving up your laundry baskets.

Example: Laura's Uniform
My NPS uniform consisted of a flat hat with a leather band—two flat hats, one for each season: a straw hat and a velvet hat, but the same leather band for both—seasonal shirts with an NPS patch on one arm, an NPS leather

CLOTHING, SHOES, AND ACCESSORIES

belt, gray-green pants for the season and hiking, brown socks, and one pair of leather hiking boots.

I held on to these uniforms for years after my time at the NPS. They stayed in the back of my closet, clean and unused, until I finally decided to go through and sort everything.

The shirts had NPS patches. I could try to remove the patches, but it might rip the shirts. Made with a polyester-blend, they did not suit me. Into the donate pile they went. Likewise, the hats, which I had inherited from other rangers, went into the donate pile.

Few know polyester pants comprise the NPS uniform. While wearable, I do not wear green, polyester-blend pants. Therefore, into the donate pile they went. A note on the pants: these were a hot ticket item, but they were not sellable. Some tall ranger could use them much easier than me trying to sell them.

A leather belt. Oh, I loved that leather belt. A perfect fit and looked western. I had a hard time giving up this item. But I could not wear it with my other clothes. As I tried to push the belt through thinner belt loops, it squeezed the belt, nearly ruining it. With a heavy heart, I dropped it into the sell basket.

The brown hiking socks. WOW, I am still using them! If I had to, I would darn or repair them. They are that good. Into the keep pile they went. The hiking boots were made of expensive Gore-Tex and leather. They were cumbersome and meant for wet conditions. They were perfect boots, but I already had comfortable boots, and these leather ones took up space, so they went into the sell pile.

Out of the seven items, only the socks were kept.

Tip C: Storage

Most homes have closets or furniture to store their clothing. If you do not have the space in your main living area, you can store clothing in an attic or basement. You should have moth-proof storage containers or clothing bags if you want to keep them for a long time. Get some cedar blocks to put into closed containers to keep insects away.

Example
When we rode motorcycles, we needed closet space for the riding jackets, long-sleeve shirts, and layered jackets. We installed shelves, poles, and heavy-duty hangers in our toyhauler garage. We created this system when we were living in the big house, but we continued using it with the tiny homes and the small casita, big garage home. We bought shelving, bars, and other items.

The Second Principle: Recycle, Repurpose, Repair, Refresh

We recycle our clothing when it is reasonable to do so. Trading clothing with friends would be great, but none of my friends are the same size. We can usually interchange hats and scarves, but everything else has a size attached.

Buying recycled or used clothing from thrift stores is an excellent way of buying at local shops. The money goes to worthwhile organizations, and you get clothing that has already lasted through one person or more's clothing cycle.

Here is a system I use for my clothing. I ask myself this whenever I buy, keep, or recycle my clothing:

- Will this clothing last a season or longer?

- Could the clothing piece be part of the capsule wardrobe?

- Can it be a layering option with clothing below or above it?

Tip A: Repurpose

Repurpose means using the item as something else. The possibilities can be as large as your imagination. Here are a few ideas:

- Pants convert to shorts. Jeans are perfect for this.

- With sewing capabilities, you can reimagine clothing into new articles, such as dresses becoming shirts or long shirts becoming dresses.

- Sew items into non-clothing items like rags, blankets, and pillow shams.

- Long-sleeve shirts can be made into short-sleeved or no-sleeve shirts.

- Scarves can be made into many items if they are big enough. I have used bandanas as belts, napkins, shades, hats, bags, and sacks.

- Long-sleeved t-shirts can be made into draft protectors. Cut off the sleeves, sew one end, and stuff the rest of the shirt inside. Sew the other end. Put your draft protector next to a door or window to keep out the winter winds.

- Oversized shirts become cover-ups for summer. You can decorate them with fabric markers and beads.

- Turn sensible shoes into dressy ones with a pin or flower.

- Turn earrings into necklaces or charm bracelets with a long or short chain.

- Turn clothes made of cotton or fleece into reusable towels to replace paper towels.

Another idea is making ornaments, artwork, or other keepsakes for the holidays.

Fabric from old shirts, especially flannel, can make reusable wrapping paper for holidays.

Example
An artist friend of mine gets clean t-shirts from thrift stores. She looks for ones with unusual designs. From those shirts, she makes kimonos. She has become well known for her unique and unusual kimonos, a perfect statement piece for a minimalistic wardrobe.

Tip B: Repair

You can repair shoes. This means to add new soles or inserts to them. Sasha did so with his Crocs using Vibram soles and Gorilla Glue. While they did not look pretty, they became more wearable.

You can also take clothes to a seamstress for repair. Patching holes in pants or blazers harks back to the scholarly Indiana Jones. Patching the knees of your favorite jeans helps them last another season. And adding a zipper or elastic material helps you keep the item you need a little while longer.

CLOTHING, SHOES, AND ACCESSORIES

Tip C: Refresh

Refresh clothing, shoes, and accessories by cleaning them with a gentle wash. Don't wash leather shoes. You make them look new with polish. Many accessories require a gentle cycle or hand washing.

- You might find a white shirt or sweater dingy, so dye it instead. The same goes for fabric shoes; try to dye them instead of throwing them out.

- You can even use craft paint or a craft machine to give plain shirts a new life with a fun design.

- You can get combs for removing the little pills or cloth balls on cashmere and fine wool sweaters. Sharp scissors work with sweatshirts too.

- You can use iron-on or sew the patches onto your favorite sweater or flannel shirt for a new look.

- My artist friend takes old jewelry pieces and makes new pieces of jewelry. Adding a pin to a blazer, shoes, or a dress makes the whole outfit pop. Costume jewelry can be remade into artwork or other types of jewelry.

Throw all clothes, shoes, and accessories for repurposing, reusing, or refreshing into the recycle basket.

Example: Laura's Uniform
Those seven items that I had from the NPS are prime examples of the recycling principle. The reason? Several of my uniform pieces were owned by someone else. My first flat hat saw at least three summers before it hit my head.

All leather items received polish, so they shined. All the shirts and pants were clean before donating. The socks I kept.

The Third Principle: Sell

As you go through your stuff, there will be a sizable pile in the sell basket. One may wonder how much money they will make on all those t-shirts and purses. Not much, I am sorry to say. Even designer purses and shoes, or that Armani suit or the latest designer dress, will not get you what you paid for it unless you bought it secondhand.

Fashion prices rarely increase. The only items I have seen are metal jewelry, such as gold necklaces, or gemstones like diamonds. If you are lucky, take care of your items, and rarely use them, you might see higher prices. Most likely, though, not at the original price.

However, if the piece is an heirloom or vintage quality and you are not the original buyer, you may get more money. If you bought it at a thrift store or consignment shop, you might recoup your purchase price if the piece was well-maintained.

Imitation and costume jewelry are especially hard to sell. Historical (1940s, 1950s, and 1960s) costume jewelry used inexpensive simulated gemstones, set in pewter, silver, nickel, or brass, and led to high volume and cheaper prices. The costume jewelry of today uses crystals, cubic zirconia-simulated diamonds, and sterling silver or silver-plated brass. However, if priced right, you can sell it all. Many crafters look to costume jewelry for the beads, to make something new.

Ten Steps of Tips: Take Pictures, Know the Worth, and Where to Sell

This step-by-step process uses three tips—take pictures, know the worth, and where to sell. It is not a long process. **Take emotions out of the equation.** Look at each piece with a critical eye. Once the feelings are gone, give yourself a treat and begin.

Look over each item with care, ensuring no stains, tears, fraying hems, pills, or discolorations exist. If you find issues, put them into the clean, refresh, or donate baskets. If it passes muster, put it into the sell basket.

Tip A: Take Pictures

Take the sell items out of the basket and prepare them for photographs. Set up the ironing board and iron, the steamer, and a white tablecloth on a table or hang on a door. Turn on all the lights in the room. If you have directional lights, use them. If the item is small, use a box under the tablecloth to get close-ups without zoom.

Start taking pictures of each item.

- Clothing: you want an overall picture, a closure (for buttons, zippers, or strings), and an unusual picture like a label, receipt, or price tag.

- Shoes: you would have the overall picture, the soles, and the inside label or inside of the shoes. Adding a picture of yourself with the shoes on may help them sell.

- Accessories like scarves, hats, and gloves: have at least two pictures—an overall and a label or price tag picture.

- Jewelry: a comprehensive picture of the overall piece and the clasp for necklaces or bracelets is critical.

After taking the pictures, take a break. Being under those lights could be a strain on your eyes and patience. Go outside for a bit.

Tip B: Know the Worth

Look on Amazon, eBay, and appraisal sites to find the worth of your stuff. The most money for your stuff is on Amazon or eBay. Getting your property appraised would make sense if you believe it to be vintage, old, or high-value. You will make the least money at pawnshops and consignment stores.

Tip C: When and Where to Sell

Determine when is the best time to sell an article. Selling winter coats or furs in summer is hard unless you live in cold climates. While selling short-sleeve dresses in winter is difficult, people sell them anytime if the dress is resort-wear or preppy.

Where you live will also increase or decrease your profit. You should check them out for the commissions and bring pictures of the best things. Follow the rules. While selling online is possible, selling at consignment shops gives you better opportunities if you live in a metropolitan area. These towns and cities will have a variety of consignment shops to sell your stuff. Here are some metro areas that stand out:

- New York pays for your old blazer. Nebraska pays in guilt.

- But really—New York Metro area: Connecticut, New Jersey, and New York.

- Historical areas in the Washington, D.C. metro area.,

- Southeast historical metropolitan areas, especially Charleston, N

CLOTHING, SHOES, AND ACCESSORIES

.C. (*Designing Women* show) and Savannah, G.A.

- Austin, TX for its young and hip vibe.

- Phoenix, AZ area, most likely Scottsdale and Paradise Valley, for western wear.

- Las Vegas, Boulder City, and Henderson for high-end clothing worn by the stars.

- San Francisco and Palm Springs, CA, for vintage and resort wear, respectively.

Online Selling: eBay, Mercari, and Poshmark are the best places to sell clothing and shoes. A second favorite is Etsy.com, as they allow vintage sales. Amazon is better for jewelry and accessories where there are SKU numbers.

- Facebook Marketplace, while online, still needs face-to-face interactions. The marketplace works for singular items or to advertise your garage sale.

- Craigslist is also online with face-to-face interactions. We find the list works well for selling other items, but not for clothing. The list works well for garage sale advertising.

- Sotheby's and Christie's sell higher-priced or appraised items. They take commissions and depend on the worth of the object.

- One of the best ways to sell your lower-cost items is through garage, yard, and estate sales. Garage sales are usually your best bet. It costs you much less upfront to hold a garage sale. Some communities have "garage" sales in a parking lot.

- One often overlooked place to sell is pawn shops. This is the final option, failing online and garage sales. While notorious for their low payouts, they can have beautiful items. A pawn shop might assume you want to buy back the item. You could get higher prices for your items once you tell them you do not. I suggest using the pawn shops with your "unverifiable" vintage or costume jewelry, purses, shoes, suits, or fashion pieces.

Example
We used every one of the above tips to sell items, except for the pawnshops. However, we bought my engagement ring and wedding bands at one in the Washington, D.C. area. A family friend worked as a gemologist and got us an excellent price for the gold bands and diamond ring.

The Fourth Principle: Donate or Dump

Donation becomes the way to go if you cannot sell your stuff, i.e., no one wants it. Many people need that business suit or professional wear at a fraction of the cost for their first interview or work wear. By already having the items in an acceptable repair and clean, you help the donation centers as well. The donation centers spend less time and money on the articles; you feel good about helping your neighbors and community.

Make sure the stuff you donate has known value. Even if you do not report these items on your taxes, having the information will help you remember where your stuff went.

Example: Laura's Uniform
The summer and winter hats sold fast to other NPS employees. The leather band and belt were the highest-priced items sold on eBay. I sold the hat box for a perfect price. But the shirts, pants, and boots did not sell. As an NPS

Ranger, I could not sell the NPS shirts to someone not in the NPS. The boots could not sell, and I kept the socks.

We donated the clean, long-and short-sleeved shirts, green pants, and boots to the Lake Mead National Recreation Area (NPS) Park Rangers. When I donated those pants to the rangers, my heart rejoiced. I saw a tall, dark, and handsome ranger wearing them the next day. The relief I felt that someone could use those clothes was worth the five years of hanging onto the clothing.

Donated clothes to thrift stores go through a process:

1. You bring them to the thrift store, where the workers sort the items.

2. Each piece of clothing gets a specific tag in the thrift store.

3. They land on the shelves and racks for a predetermined time.

4. A sale starts if clothes remain unsold for a few days.

5. If the clothing does not sell at the thrift store, the thrift store combines the unwanted items with similar items and sells them in lots to a wholesale company.

6. If the clothing does not sell, workers pack it with similar materials in each ton and shred it to make recycled clothing.

I believe people should not throw clothing away. However, when the stains or tears mean no prospect of using it as a rag, then you could throw it away. If the stains are bio-stained, then they need to go into a bio-dumpster, usually found at Emergency Medical Services (EMS), hospitals, and doctor offices.

Tip A: How Long To Decompose?

Clothing in landfills takes a certain amount of time to break down:

- Pure cotton gloves: three months

- Wool sock: one to five years, and the itch stays around forever.

- Leather shoes: twenty-five to fifty years, and the thrown-away polish becomes hockey pucks in metal containers.

- Rubber boot soles: fifty to eighty years

- A dress bought at H&M: six-hundred years (but they have a sustainability policy).

From one of my readers: It is okay to acknowledge when a piece of clothing has served its purpose, and knowing when to let go of an item can be just as important as giving it a full course of life.

Let us work together not to throw out a single item in this category. Landfills are overcrowded, and items decompose at slow rates.

Extra: Capsule Clothing

The capsule clothing idea is popular today, but it started during the 1970s with Susie Faux. Donna Karan popularized capsule clothing with her Seven Easy Pieces idea. In 1980, the Color Me Beautiful book came out. It incorporated capsule clothing with choosing colors of clothing which best matched your skin tone and hair color. Ms. Faux had career women in mind when she created the capsule clothing concept. The core pieces were a jacket, a skirt, trousers, a blouse, a sweater, tights or stockings, shoes, a coat, a dress, a bag, and a belt.

She stated, "The ideal size of your capsule will depend from person to person. There are no hard and fast rules here—the principle is that less is more so really what you're trying to do is to make the most of your budget to create a working wardrobe with high-quality clothes that will be sufficient for your lifestyle."

Each decade a new blog post comes out, suggesting the capsule concept as something new. *In Style* magazine even had capsule clothing as its prime article every month.

While I no longer work in a corporate environment, the capsule clothing concept still intrigues me when I look over my clothing every year. When you are downsizing, you might find the concept works well for you, too.

How To Organize Your Stuff

If you live somewhere with three or four seasons, realize you might have more clothing than those who live in a two-season area. Buy see-through, 36-inch by 24-inch by 12-inch containers with snapping lids. Put your off-season clothing into those containers along with small cedar blocks to keep moths away. Store the containers on shelving in the garage. Buy open shelving units without weight constrictions. California Closets and other shelving unit types have white wire and regular wire shelving and poles for your hanging clothing. If the place you store extraneous items gets dusty, use hanging luggage bags for your most delicate items.

Homes without closets: Invest in Scandinavian-type furniture with genuine wood, clean lines, and drawers under a platform bed. Wardrobes with enough hanging space for your pants and dresses have drawers underneath to store clothing and accessories.

One-bedroom homes with with closet: Storage may be tight with two people. Invest in bureaus for each person. Lacking space? Get a floor-to-ceiling wardrobe; buy or build it. Put seasonal clothing on the upper shelves or drawers and rotate for the seasons.

Wear clothing multiple times in a week unless they get dirty. Wash once a week in a washing machine without agitators, so that your clothing lasts longer. Dry in a dryer or hang dry. Use a clothing bag for delicate clothing or small items.

Two-or-more-bedroom home with closet(s): Depending on how many people live in the home, you might designate one bedroom for all the seasonal accessories such as hats, scarves, and gloves. Each bedroom should have a bureau for easy to lose clothing. Each person who uses the bedroom should have a bedside table with drawers to place their items for sleep. Having an electric outlet on a shelf above eyesight lets you charge electronics without the blue light glare.

Garage space: If your home has garage space available, build shelving going to the ceiling. Make sure the shelving is at least five inches deep. If you have space in your garage, basement, or attic, invest in free-standing shelves and hanging areas. Minimize the space you need.

How To Organize Your Clothing, Shoes, & Accessories

Here are some suggestions for the worksheets below.

Keep	Six tops with jeans	Closet or Wardrobe
Keep	Black outfit for funerals & weddings	Closet
Keep	Wool Sweaters	Switch with Summer clothes in see-through box with cedar blocks under the bed/attic
Recycle	Canvas pants	Repair or take to dry cleaning seamstress there.
Recycle	Beloved T-Shirt	Take picture of front, make graphic, use as a dust cloth or rag
Sell	Black suit	Consignment $5
Sell	Cute shoes that hurt	Clean & sell on Mercari or Poshmark
Donate	Prom Dress	Dress for Success
Donate	Winter Coats	Homeless shelters or OneWarmCoat.org
Dump	Unwearable Clothing	Any H&M store or Goodwill, Salvation Army, & Terracycle.com
Recycle	Sneakers	GotSneakers.com, DSW, & Nike

You might be able to donate / dump unused fabrics to fabric stores but call first. Same with community colleges and schools.

Organize Your Clothing, Shoes, & Accessories				
Keep		Recycle		
Item	Location	Item	Location	Repurpose, Repair, Refresh

CLOTHING, SHOES, AND ACCESSORIES

Organize Your Clothing, Shoes, & Accessories			
Sell		Sell	
Item	Location	Item	Location

Organize Your Clothing, Shoes, & Accessories				
Donate		Dump		
Item	Location	Item	Location	H&M or other

CLOTHING, SHOES, AND ACCESSORIES 69

You can find these tables by signing up for my email newsletter at LauraJevtich.com/OYSV2 and you can print these out over and over.

Household Goods

Applying the Four Principles to Your Kitchen and Linens

This is a big category, which includes dining goods, kitchen goods, bedroom linens, and bathroom linens. Most of these items are durable goods that you use on a daily or weekly basis. Some stuff only comes out at particular times, like holidays or birthdays. Other times, they decorate your home daily, monthly, or yearly.

Excessive home possessions trigger stress and anxiety. "Keeping up with the Joneses" is a flawed concept. Not only does the excessive stuff harm you in a mental capacity—it also initiates memories. While memories are important, they need not control your life. Taking pictures of the stuff you owned during joyful events allows you to bring back those memories without the stuff.

As Sasha states it: "Do not let your stuff own you; you own your stuff!"

Example
I never got into decorations for the big house. I had wall artwork and gardens outside. Our furniture and home design focused on clean lines, less clutter, and keeping everything as dust-free as possible. This concept spilled over into our tiny homes and small casita.

HOUSEHOLD GOODS

I am not a person who gets enjoyment from knick-knacks. I grew up in a household with those items, and one of my jobs as a child was to dust. Dusting every curvy nook and cranny of small objects was not my idea of fun. After a traumatic shipping event, where I subconsciously fell on a box of porcelain plates, ten years of dusting were gone in a glorious crash, and my back was wrenched.

What Is In This Category?

Dining and Kitchen Goods:

- Plates, cups, glasses, and mugs

- The wares: silver, china, and tupper

- Spoons, forks, knives, and serving utensils

- Crystal

- Pitchers

- Bowls and serving platters

- Soft products such as linen napkins, tablecloths, and placemats

Small Appliances:

- Mixer, food processor, and pressure cooker

- Blender, coffee maker, and kettles

- Crock pot

- Rice, pasta, and bread makers

Cooking Tools:

- Knives
- Mixing bowls
- Pots and pans
- Muffin or bread pans
- Cookie sheets
- Salad spinner
- Graters

Extra Items:

- First aid kit
- Fire extinguisher
- Vacuums, brooms, and mops

Linens:

- Bedroom sheets and pillowcases
- Pillows
- Comforters
- Blankets
- Bathroom towels, washcloths, and hand towels
- Beach towels

A comprehensive list helps prioritize tasks.

The First Principle: Keep

This broad category reveals ownership patterns skewed toward a surplus of certain items and a shortage of necessities. During the winter or summer, you will see grocery stores filled with the "latest and greatest" fashions for your home. Don't buy into the hype or promotions. Having five patterned china sets only means more handwashing. Case in point, how many plates do you need for a family of four? Or two?

The category contains varied items. Start with one kind or one group of stuff.

Tip A: Similar Together

To make this process easier, start in your kitchen. Get out all the dishes and sort them by pattern or type. This means keeping all the plates in the same area, all the bowls together, and all the smaller plates neat and stacked. For now, put them on your sorting table. You will decide later on whether to keep them.

Ask yourself these questions to decide about your stuff:

1. Take measurements in cabinets to find space for all the dishes (or whatever the category is).

2. Are they accessible?

3. Are there too many of these items? Too many means you must be creative to store the items.

4. Do you need all of them? Or do you want all of them? This

question, in particular, is the emotional one. Try to figure out how many you've used these items over the past year.

Example

In almost every home I've lived in, I've made sure to measure the dimensions of the cabinets. This allowed me to make a spreadsheet or list for each cabinet. Upon inquiry regarding the item's location, I consulted the list. Each cabinet in the kitchen, no matter in the big home, tiny homes, or casita, I knew the item's location. Extra storage, whether food or household goods, went to taller cabinets.

I used to have an RVing friend who could easily have been featured in the Minimalism Hall of Fame. He owned one cast iron skillet, a sharp knife for bread and tomatoes, and a pot that burned nothing. The rest? He's still single.

Tip B: Usefulness

How often have you used that dish or item? You want to take each piece, each type, or each set and figure out its usefulness.

Have you or your family used it today?

↳ If not, have you used it in the past week?

↳ If not, have you used it in the past month?

↳ If not, have you used it in the past three months?

↳ If not, have you used it in the past six months?

↳ If not, have you used it in the past year?

Why are you holding onto it? If you have not used it in the past year, there may be sentimental reasons for keeping it. First, understand that feeling is normal, then remove those emotions.

Remember, take your emotions out of the formula for paring down your stuff. It will be hard at first. You may realize you have too much when you see all of your things on the table or floor.

Example: Laura's Household Stuff
One type of houseware I had a lot of was my grandmother's silverware. They're made of real silver, meaning you must polish and take care of it. When she died, I got her set. I loved the clean, straight lines and wheat symbol on the handle. When we married, the family heirloom became my formal silverware.

When I received the set, it came with eight place settings, as well as serving and carving pieces. Over the next eight years, I continued to add to the place settings. I kept the silverware in a beautiful Art Deco cabinet covered with cloth. All the silver was in the same cabinet, although in separate drawers, as it did not fit into one.

When we decided to sell the house and everything in it, there were twenty place settings, plus extra spoons!

Were these twenty place settings helpful to us in an RV? Would we have enough storage available? I decided, after much soul-searching, to split the set. I kept four place settings and gave the rest to my sister. She has two children and will provide them with the silver to keep it in the family. The set functions for two families.

Tip C: Containers and Storage

You have cabinets and drawers in your kitchen. You might also have closets or pantries for your linens and cookware. Here are a few suggestions on keeping your dishes, housewares, and other items chip-free yet accessible.

1. Invest in silverware holders for valuable silver; you are polishing without even doing so.

2. Flatware organizers work well in a large drawer holding all the utensils used in a kitchen.

3. Instead of stacking your plates, invest in plate holders to reduce chips or scratches.

4. Glassware openings should point down. Why, it is usually more space saving to alternate.

5. Find room for wine glasses to hang by their stems. Or have wine glasses without stems.

6. If you are in an RV, use reusable plasticware rather than glassware. A moving vehicle breaks those hanging glasses. I know, as I broke several before I learned my lesson.

7. Invest in a lazy Susan. This is a type of tray for the inside of your cabinets. Some cabinets are quite deep. Accessing the rear of the cabinet presents a challenge. A lazy Susan allows you to turn it and get the things that are hard to reach.

8. A knife rack will help keep your knives from dulling. If there is no space for one, cover each sharp blade with a cardboard sleeve to prevent dullness and avoid slicing your fingers.

9. Have ONE junk drawer in the kitchen. This might hold the extraneous items you need.

10. Keep a drawer or a container for your table linens. Table linens can go in a linen closet, bathroom, or in an outside kitchen. Make sure whatever you use has easy access.

11. Keep cords for your small appliances wrapped when unused, and put them away in a cabinet or pantry. This keeps them neater.

12. Use your linen closet for the linens in the surrounding area. If you use linens outside, have a drawer or shelf for them, and keep them clean.

13. Keep linens in a moth-proof, zippered, vacuum-packed bag. This is a must in an RV.

14. When possible, hang large tablecloths in a long closet. Use a resealable bag. They stay clean and require very little ironing.

15. Use paper bags for some of your cookware, including your crock pot, when in an RV. Since the pot portion is ceramic and the base is metal, wrap the ceramic pot in a paper bag. This prevents the pot from chipping while moving.

16. Use a container, small tub, or drawer to have all your plastic reusable containers together. Any containers without tops go to the donation box.

17. Use small or stacking shelves on your counter to give you more space. Use the cabinets' shelves for maximum usage if they are moveable to different heights.

Example

We had an upstairs linen closet in the big house. Each shelf held one size of bed linen. One shelf was for the towels, and one was for the comforters or blankets. We only kept bedroom linens in that closet. In our tiny homes, we stored linens in the bathroom closet, and we kept the bulky winter duvet in a drawer in our bed until we needed it.

In our tiny homes, each container or small appliance had its spot underneath the kitchen sink. We donated any container that didn't have a lid.

In our casita home, I try to keep four sets of king-size sheets—two winter sets and two summer sets; the unused linens go in a vacuum-sealed bag. We keep the outside linens in a storage cabinet meant for outdoor use.

Many options exist for retaining possessions; however, you may wish to recycle, sell, or donate instead.

The Second Principle: Recycle, Reduce, Reuse, Repurpose

There are plenty of ways to recycle your household goods without putting them into the trash. Once you have decided what you will keep, recycling comes next, with a few tips on recycling the stuff you do not want or need.

Tip A: Reduce

As I did with my silverware, you might decide you do not need something. Two diners require far fewer than twenty place settings.

"Reduce" can also mean not having an item to begin with. If you have a large gathering, ask friends if you can borrow their dishes or glassware.

Rent when you want consistency. Renting allows you to use but not maintain and store it.

This is a hard recycle tip. While you want to keep items, sometimes the items themselves end up overwhelming you, or the items you store get too heavy to lift. Why not reduce the amount in a set by passing some of them on to new couples starting their homes?

Tip B: Reuse

Reuse your plastic containers to pack lunches for work rather than using disposable products. Likewise, use linen or cotton tablecloths rather than plastic. People use tablecloths multiple times, and they help the environment and look nice. A tablecloth only needs a "whack, whack, whack" to get clean, and your beautiful table will not get scratched.

Reuse also means buying and using secondhand. Thrift stores usually have many of the small appliances used in the kitchen.

Tip C: Repurpose

Repurpose means to use your stuff in a new way. Your salad spinner's handle broke, but you can repurpose it as a serving bowl. Your crock pot cooks rice to fluffy perfection. Your food processor makes dough, mixes cake batter, and slices food. Why do you need a bread maker, a mixer, and a slicer? Minimize the amount of appliances you need and repurpose the one for multiple uses.

We use our steak knives as paring or utility knives. Our dish towels become rags when they no longer soak up the water. Tablecloths become placemats, which become napkins and then rags or patches for clothing.

The only limiting factor to repurposing an item is your imagination. Here are some other suggestions:

- Tall cups can be used as toothbrush holders.

- Baskets can hold potted plants.

- Serve potato chips on a linen napkin in a wicker basket.

- Picture frames can become serving platters; paint or decorate for each holiday celebrated.

- Plasticware can be used for medicines, food, or crafts. Have clean plasticware.

Example

We have a Mikasa dinnerware set in our casita home, in the kitchen cabinet. I bought the collection at a thrift store, which reduced the price because the china was not a full set. There are eleven plates, eight dessert plates, ten soup bowls, and other pieces. One plate has a slight chip on the underside of the lip.

After emptying the dishwasher every Monday, I take two plates from the top of the dish stack and place them below it. I do the same for the two sets of three glasses, changing them weekly. The same is true for the sharp steak knives in the knife block.

Depending on how many of that one item you have, you could do the same type of turnover for your plates, glasses, or silverware for a month. They are all clean and go away clean, but regularly using an item can make it break, chip, or fall apart if you do not use this recycling method.

Rotate dishes like tires—even wear means they'll outlive your relationships.

The Third Principle: Sell

The household goods category is large, and your stuff will have diverse price points. If the items are vintage, heirloom, well taken care of products without pits, stains, chips, or missing pieces, they could fetch higher prices. A blue glass collection, near perfection, will bring in more money than the latest Disney characters on plastic glasses from a Happy Meal. Because of its functionality and limited uses, a fondue pot will bring in less than a crock pot, and a food processor will never sell at the price bought new.

Household goods are always numerous. However, if there are complete sets of chinaware, crystal, and silverware, those could bring in more money as a set than individually. When you decide to sell your stuff, try putting items into sets. Heirlooms sell better as sets. Your grandma's china? Worth more together than scattered across eBay.

Sometimes, and depending on your situation, selling part of a set will work better than selling the whole thing. Make your decision about splitting the set before you sell.

Example
As I stated earlier, I learned a lot by watching my parents, aunt, and uncle run estate sales when I was growing up. An estate sale usually occurs after the owners have passed away and the heirs wanted to sell the property. None of the sales they did were auctions. They were estate sales featuring the contents of the house, garage, and other buildings on the property. They usually lasted four days—Thursday through Sunday. High-end jewelry, silverware, or gold pieces were sent to be appraised and then to auction houses.

Tip A: Take Pictures

While most items can go on a white background, white and clear dishes look best on black tablecloths. Make sure those tablecloths have no pills or hair. Silver and gold look best on black tablecloths.

If you have the room, separate the sets and place them on a long table with a black tablecloth.

Take pictures of the set:

- Use directional lighting, and do not keep any pictures with a throwback flash.

- Take pictures of the inside and outside of pots and pans.

- Show a tape measure of the length of the tablecloths, as well as for sheets and towels.

- Measure the dishes and bakeware and put cards next to each set so that other people do not have to ask or guess the sizes.

Tip B: Figuring Out the Worth

A fancy chinaware dish you bought for $25 might now cost $3. A place setting of dinner plates, salad plates, soup bowls, and mugs bought for $50 might go for $10 now. As soon as you purchase the item, its value decreases. What sells in retail stores is not the same in estate or garage sales. You likely will never recover the cost.

However, if you take care of your things, you might get one-fifth of the price. Again—especially in sets. Linens will never be the same price, usually going for one-tenth to one-twenty-fifth the cost.

Where to find prices? Look to Replacements.com for dinnerware, silverware, crystal, and other high-end household goods and see how much they're going for. Look at eBay listings for the prices sold. Amazon is best to use for small appliance prices. Check prices at antique stores and consignment shops if you have the time and inclination.

Tip C: Where to Sell

Do not send china, glassware, or other breakable items off site. Shipping items to sell can be difficult with breakage and high shipping costs. Too often what you send will break, and you will not get the total price. If you want to sell piece by piece, the best way is by consignment or through antique stores. Have your pictures ready to show before you bring your items there.

Otherwise, having a house sale, estate sale, or garage sale is the best way to sell household goods. Most items are too heavy for shipping. Craigslist is another option if you do not want people to visit your house. You can go to a centralized location instead. Facebook Marketplace might be an option, especially in a large community, but beware of the scams and the tire-kicking folks asking you for lower prices.

You can try setting up at a flea market or community garage sale. While there is a setup cost, you might find more people at these events. You will find more buyers for your stuff through these avenues. You will have more competition, but more people might see your goods.

Do not sell breakable items that need to be shipped. The possibility of broken things means lower ratings and lost money.

The Fourth Principle: Donate or Dump, Rethink, How Long to Decompose, Resources

Tip A: Donate

If you have visited the thrift store lately, don't be surprised to see all the mismatched household goods. Anything that does not sell can go to a thrift store. My parents' TLC business would make trips to the thrift stores after each sale. However, stained or ripped linens are better used as rags than donated items.

Look below for some simple ideas on rethinking your choices, how long your stuff takes to decompose, and where you can go to recycle your stuff.

Tip B: Rethink

Landfills are unnecessary for this category.

- Broken glass can be art... if you're an artist. Otherwise, stop hoarding shards and recycle them with your glass bottles.
- If you *are* an artist, try making broken dishes into tile mosaics.
- Use linens and fabrics as rags or stuffing for crafts.
- Use plasticware or plastic containers to store things.
- If you break the crock pot insert, pick up another at a thrift store or garage sale. Use the broken pieces for grilling over a fire or for artwork.
- Losing a part of a small appliance does not mean the item is useless; you can repurpose it or call or look up the appliance maker.

Sometimes, they will send you the replacement for free or at a small cost.

Tip C: How Long to Decompose?

This is how long common materials take to decompose in a landfill:

- A paper bag takes one month to decompose.

- Cardboard takes two months to decompose.

- Plywood takes one to three years to decompose.

- Steel takes fifty years to decompose.

- That plastic spoon from the local fast-food joint takes two hundred years to decompose, and some don't decompose at all!

- Aluminum takes anywhere between two hundred and five hundred years to decompose.

Plastic and Glass Items
Plastic is particularly long-lasting. A plastic container takes between twenty to five hundred years to decompose. Plastic bags take twenty years, plastic-lined coffee cups take thirty years, plastic straws take two hundred years, plastic bottles take four hundred and fifty years, disposable diapers take five hundred years, and fishing lines take six hundred years.

Glass can take four thousand to one million years to decompose in the environment, and even longer in landfills. Archaeologists have found human-created glass in Mesopotamia that is thousands of years old and still has not decayed.

Tip D: Resources

How to recycle small appliances:

- Make sure that it cannot be repaired.

- Remove the battery and recycle the battery.

- Dismantle the appliance to harvest parts or components for reuse.

- Some items can be shredded. When shredded, plastic, glass, and metal materials need separation from each other. Manufacturers subsequently receive these items for new products.

There are buy-back programs for small appliances from Walmart and sometimes local stores, where you can receive a gift card for your items. Different stores have different policies, but you might also try your local hardware stores.

Look on Pinterest or other sites to learn how to use your broken pieces as art. Mixed media has several uses. You could even contact an art club, quilting group, or school to see if they want your scraps and broken pieces for their next project.

How To Organize Your Stuff

Household items seem to accumulate without pause in most homes. You go to a party and bring back a plate of leftovers, and you forget to return the plate. Or your casserole dish stayed at your friend's house for a month before you remembered who had it last.

Kitchens without a pantry or closet: Most galley-type kitchens are notorious for their lack of space, but they usually offer a surprising amount of storage. Invest in stackable shelf organizers—these decrease the height inside the cabinet, allowing for more storage.

Unless you have parties every week, you should only have twice the household goods to those living in the home. For example, two people live in the home, and you have four mugs. This makes sense, as you can use the mugs for hot drinks and for soup or chili. Having over four mugs becomes problematic, and where can you store them all?

Kitchens with a pantry or closet: You need to decide if you are going to have a step stool for those hard-to-reach items at the top of the cabinets. Use sturdy ladders for items stored above two feet. Pantries with depth allow you to use the back of the door for spices or narrow items. Put heavy small appliances on the sides of the pantry while the food stuff is in the center and easy to access. Use logical places for dishes and glasses, such as close to the sink or dishwasher.

Garage space: If your home has garage space, a basement, or an attic available, build shelving going to the ceiling and use it to store seasonal items. Make sure the shelving is at least five inches deep in order to fit seasonal plates and dishes. If you have a favorite dining set for the holidays, protect it with quilted storage containers and thin linen napkins between the dishes. The best idea? Have the linen napkins and dishes match for the holiday or celebration.

How To Organize Your Household Items

Here are some suggestions for the worksheets below.

Keep	Matching wine glasses	Cabinet
Keep	Tomato Knife	Knife Block
Keep	Mismatched China	Separate out the different types, keep most similar together and sell the rest.
Recycle	Crystal Vase	Repurpose Art
Recycle	Bits of doodads or spare parts	Wearable Art
Sell	Bread & Pasta Makers	OfferUp / $20
Donate	Whatever did not sell	Local Thrift Store
Dump	Bits & Shards	Arts district or Community College

You can find these tables by signing up for my email newsletter at LauraJevtich.com/OYSV2 and you can print these out over and over.

Organize Your Household Goods					
Keep		Recycle		
Item	Location	Item	Location	Refresh, Repurpose, Reuse

Organize Your Household Goods			
Sell		Sell	
Item	Location	Item	Location

Organize Your Household Goods				
Donate		Dump		
Item	Location	Item	Location	Arts or other

Physical Media

Applying the Four Principles to Books, Maps, Photo Albums, Scrapbooks, Music, Videos, and Documents

When asked, "If you could only take one book with you to a deserted island, which one would you take?" My first thought is—I have so many wonderful books, how could I choose? Maybe I'd pick Ten Steps to Escape a Deserted Island - And Ditch Ships Forever!

When we lived in the big house, our bookcases and rooms overflowed with books, each one like a treasured friend. We also had boxes full of photographs of our nieces and nephews, photo albums full of Sasha's partying days, scrapbooks filled with cute paper and embellishments, and our wedding albums. Plus, we had a filing cabinet for my business' essential papers and documents.

Sasha's music collection is legendary. The number of albums and songs is more extensive than most XM channels. He has a decades-long collection that he started when he was eight or nine, and it continues to grow. We also have an impressive video collection. This category and chapter will explain how to digitize most media for small spaces.

The First Principle: Keep

Your collections are growing, and you have no control! You find stacks of books, papers, and magazines all over the house. Any time you need to find something, it takes hours of digging and searching through the chaos. Take a deep breath, get your baskets, and let's get organized.

Tip A: Put Everything Similar Together

Start with the smallest room in your house and clear off the table. You will have many things to look at and decide about, so let's get comfy.

Put all the books together in one spot on the table. Do the same with maps, notebooks, photographs, photo albums, music, videos, and documents. Use the space for your other stuff if you have nothing in one section.

Once everything in this category is on the table, you can decide whether to keep, recycle, sell, donate, or dump. However, the dump will not be a trash bag but a basket. We will get to each of these sections shortly. Knowing what you want to keep gives you a better idea of how to store it later. If space is at a premium, you must be firm in your decisions.

Tip B: Usefulness

Go through each of the items you've placed in the keep basket and decide if you really need them. This should become easier for you as you go through the category. However, you still need to check your emotions at the door.

Have you used it in the past week?

↪ If not, have you used it in the past month?

↪ If not, have you used it in the past six months?

↪ If not, have you used it in the past year? If not, is it special? Or is it meant for a unique area or part of your life?

Do you want to keep it if you have not used it, looked at it, listened to it, or watched it in the past year? You and your family need to ask and answer some hard questions.

Example: Our Photographs
Sasha had a lot of photo albums from high school, college, and the time before me. I just had a few scrapbooks. These had sentimental attachments, especially our wedding albums. We wanted to keep the photos but ditch the albums, so we decided to digitize the photos. The albums were placed into the recycle basket.

If you want to keep the bulky stuff like photo albums, records, CDs, Blu-rays, DVDs, books, and maps because they have sentimental value, you can—and you can save space by converting them to a digital format.

If you are not sure, ask yourself these questions:

- Do I remember or know where these photos were taken?

- Do I know who these people are and what we were doing?

- Did the artist sign this music album or book?

- Is it possible to digitize it? If not, could someone else hold it or use it?

- Can I use this map, book, etc. again? (Be brutally honest here.)

- Do I need it for my taxes, the sale of the house, or personal property? (These would be legal documents.)

- Do I need this for my clients or family? (Such as contracts, wills, etc.)

- Do I need this for myself or my family for our health? (Such as X-rays and medical information.)

- Is this media something that brings me joy every time I see it?

Once you have answered these questions about your stuff, then you can move on.

Tip C: Containers and Storage

Have you gone to the library lately? We try to go to them weekly or monthly, depending on our location. They have shelves and shelves of books, bins upon bins of music and movies, and hanging racks for maps and newspapers. Why not do something similar in your home? If space is a premium, and all you have is a small bookshelf, a little creativity might be in order.

Walmart and IKEA both have great ideas for storage. We had several bookcases in our RV, and we had one cabinet where we stored all of our maps. Now we have a floor-to-ceiling bookcase in our tiny casita.

To prevent books from falling while we traveled in our RV, we employed a tension cord system. This way, there were two rows of books on the TV shelf. They did not slide during movement.

For those of you in a house, you have better options and more choices. You can elect to display your collections or not.

Other options:

- Milk crates, bookcases with shelves, drawers, and filing cabinets

- Magazine racks

- CD/DVD folios

- Map racks with frames to hold more extensive media.

- Filing cabinets, an accordion file folder, or a safe

For those in a tiny home or small space like an RV, having a fireproof, waterproof safe is a brilliant investment. Your essential papers, external hard drives, and valuables go there. The filing cabinet would be for the nonessential papers. We have an accordion file folder inside a safe for the essential documents.

The Second Principle: Recycle, Reduce, Digitize

The recycling part of this chapter is long, because there are so many types of physical media in this book. The category could be huge and paring down can be quick once you have everything sorted as stated above.

Tip A: Reduce

What reduce means when discussing recycling: Reduce the amount of stuff you have or will have in the future. Buy durable products that can withstand pressure, wear, and damage. Think of your stuff this way: before bringing the media into your home, can you borrow it? Can it be digitized and not take up space? Let us go through the types.

Books, magazines, and newspapers

Yes, I know you love the smell and touch of the written word. You love getting newspaper smudges on your fingers, table, or clothing. You love learning about your community and its events in the newspaper. What if you got the information online?

- Put the media on your Kindle, smartphone, or digital application.

- Several major newspapers have applications to read your subscriptions. Your local newspaper might be online.

- Apple News, Yahoo News, and Google News work well to read the news.

- Email newsletters (e-newsletters) allow you to find stories and information on a variety of subjects. **Sign up for my e-newsletter at LauraJevtich.com** to receive helpful hints on organizing your stuff.

- If you have a physical newspaper or magazine, ensure you can recycle it later.

Maps

Back in the day, all maps were paper, and you could get them at any visitor center. Now you can use applications such as GAIA, a Global Positioning System (GPS) application, which is also an offline application. GPS chips are included in most smartphones. If your tablet has cell service, it will also have a GPS chip. Examples include Google Maps, Apple Maps, AVENZA, and other applications.

- If you are in a particular area for a while, you can continue to use your old maps. However, if the town you are in keeps changing, update either to digital or new maps.

- Recycle your older maps and use them for craft projects, or donate them to a local school or library.

Photographs, photo albums, and scrapbooks

Despite the photography world's dramatic evolution, we will always look at pictures and photographs. You can store your photos by digitizing your pictures and using smartphones. Software programs for creating digital scrapbooks are available online or offline. For those who want to have physical photo albums or scrapbooks, as you consider them your craft or creative avenue, do not despair. Use your to create photo books and calendars. Digitizing also means you can make a copy to store in a cloud or external hard drive. Never lose them to fire, flood, or sticky fingers again. Also, consider a digital picture frame—it will rotate through photos, and every time you glance over, another memory is there.

- Companies like Shutterfly, Blurb, and Apple can create photo books for you.
- Zazzle, RedBubble, Spreadshirt, and Amazon have creative shops to make your Best Grandpa mug or t-shirt.

Music and videos

The best way to reduce in this category is not to buy the hard copies. Stream or download the digital copy for later use. Or rent via subscription services from Redbox, Netflix, or Hulu.

Important documents

These are your banking, medical, financial, calendars, appointments, contacts, meetings, and passwords. Any paper statements that you get through the mail can be digitized.

- Ask for online statements from your banks, credit unions, and investment firms.

- Ask your employer to use direct deposit rather than send checks. Try not to write checks. Use a debit card or credit card, or bill pay through your bank.

- You can check medical information, such as X-rays and test results, by logging onto the medical portals.

- Calendars, appointments, and meetings can be kept on your smartphone or planner. Your address book is your contacts on your smartphone. Business cards can be scanned using software or your phone. I had a business card scanner when I was running my business. However, now you can use your smartphone as a scanner for business cards.

Tip B: Digitize

Many of the items in this category can be digitized. You need storage, time, and the knowledge to do it. Make sure to scan at the highest level you can afford. Your future self will build a shrine for you.

Example: Sasha's Music Timeline

1965: Beatles vinyl

1975: Laserdiscs of famous concerts

1985: Mixed tapes for beach parties and crushes

1995: CD shelving lovingly handmade by his Dad

2009: Customized and engraved iPod full of MP3s

2024: Personal cloud drive for home and travel

Books, magazines, and news

Ebooks, Kindle, Nook, Google Play, and other e-readers exist for a reason. I bought Amazon Prime for reading and the books. I got a Kindle to read more books than I could store in an RV or tiny home. Just like Sasha's collection of music, my collection of books is massive. I can usually read two or three books at a time.

Example
We purged our collections when we sold the big house and everything in it. Ebooks on the Kindle started becoming popular around that time, but the equipment was expensive. It was hard for me, as many of my books I considered friends, rereading them each year.

Eventually I sold my books and digitized them by buying a Kindle. However, I kept many of my favorite books by giving them to other family members who would enjoy them. I also found a secondhand bookstore in my area to take back books I finished and got credit for ones I wanted.

Nowadays, you can get ebooks to read on your e-reader, like a Kindle. Or you can read books on a smartphone, tablet, or computer. Amazon's policy is that you can send the book to your smartphone, Kindle, tablet, and laptop via the Amazon cloud.

Some classic books are available to download free as a portable digital file (PDF). You can read magazines online, through subscriptions or applications. We subscribe to newspapers like the Washington Post or the New York Times and read them on our phones, laptops, or desktops.

Of course, you could go to the library. They have so many books, magazines, and newspapers. Some libraries even have e-readers on loan.

Are your children having problems reading? Get an e-reader with speakers so they can listen and read along to audiobooks. By seeing and hearing the words, children learn faster. Adults listen to audiobooks as well. At your local library, they offer free resources, such as Hoopla and Libby. You're bound to come across a great story or two. My memoir, *Unusual Adventures Traveling Through Life*, is available on Hoopla.

Maps

You can digitize maps through a software program. Maps are large, and your phone or tablet is tiny in comparison. Sometimes, the best method is to have a paper map. Otherwise, having a GPS or Garmin helps you get where you need to go, and so will apps Google and Apple Maps.

If your GPS directions turn you on to a dirt or gravel road you did not expect, turn around immediately. This might be the case with bad Wi-Fi or cellular service.

Death by GPS or digital maps is more common than you think.

Example
Death by GPS is a term used by Death Valley National Park rangers, coined in 2009 when a woman and her six-year-old child turned off a paved road onto a dirt one. She was following directions from the GPS, which led her off the road and into sand. Instead of using her intuition and logic, she continued driving and got stuck. With no cellular service and little water, she and her child started walking out. Tragically, her poor decisions led to the death of her child.

I heard of the incident while working for the NPS at Cape Hatteras National Seashore.

Photographs, pictures, and scrapbooks

Example

At the time of purging and digitizing, a large scrapbook store called Recollections was down the street. I took our albums there and scanned the photos and pages for $1 a page. I scanned them at the highest resolution possible, and we stored the pictures on an external drive.

Later, we put the pictures on a computer and a larger external hard drive. When we had the big house, our digital system was a terabyte, very large by 2006 standards, although it is puny now. Our casita home has close to twenty terabytes on Network Attached Storage (NAS), while the desktops have five terabytes each. While some might suggest storing them in the cloud, we do not always have reliable internet, so we keep them at home.

To do the same, use a top-loading scanner, a smartphone, or go to a shop like CVS, Walmart, or FedEx to scan your pictures. Some libraries also allow you to use their scanners for lower prices.

Most often, pictures and photographs are digital from the start. Store them on an external hard drive or in the cloud. Give them enough information. While most have the GPS, naming the files will give you more information. That information will help you understand the what, who, and how questions.

A friend of mine stored all of her photo albums in the basement. One night, the river close to her home crested because of the spring snowmelt. Her entire town was a disaster zone, and she could not get back into the home until three days later. Her wedding albums and children's picture albums were a mess of gooey slime and mud, destroyed forever. Thankfully, we had pictures of her wedding and the kids stored on our external hard drives. She made digital scrapbooks of the pictures, which she pulls out at birthdays and anniversaries.

Music and videos

While today's world is all about streaming music and video, this was not always the case. Sasha started collecting music in 1965, and film and video in 1968. Back then, you owned the media, not rented it for a while.

Before we sold the big house, Sasha reverted his music collection from cassette tapes to compact discs (CDs). The way he converted the first time was to buy all the music that was on his tapes in the CD format!

Thankfully, most of his spending took place before I met him.

Once the technology caught up, Sasha converted the music from the CD to an MP3 file, making them digital. The same work happened with his VHS and laser disc collections. He bought the James Bond collection at least three times: VHS, laserdisc, and DVD. He used iTunes and his computer to rip the movies into digital files. These files are stored on the NAS along with his music collection, pictures, and digital documents.

Important documents

Get an external hard drive with at least a few terabytes or more space.

You can complete your taxes online with password-protected PDFs. The IRS suggests keeping your returns or tax files for the last seven years, but you can keep them forever when digital.

Contracts, wills, and health and medical care documents need scanning and placed on an external hard drive. Typically, these documents are digital, except for your signature, and you can create a digital signature. Backup your computer once a week if not daily.

If your laptop is used primarily for documents and pictures, you can get away with half a terabyte, but why not go with a full one instead? Ultimately, you will grow into it.

Tip C: Recycle

Once you have gotten through and digitized all of your media in this category, it's time to recycle.

Here is the list of items to recycle:

1. Books: go to the Third Principle (sell) or Fourth Principle (donate).

2. Magazines: if rare or old, go to the Third Principle (sell). If it is recent and popular, go to the Fourth Principle (donate).

3. Newspapers: tie up with string and recycle that thing!

4. Maps: go to the Third Principle (sell) or Second Principle (recycle).

5. Photos and scrapbooks: go to the Third Principle (sell) or Fourth Principle (donate).

6. Music and videos: go to the Third Principle (sell) or Fourth Principle (donate).

7. Important documents: shred Fred!

Books, magazines, and newspapers:

Separate the books by this method:

1. Fabric: take the book apart and use it as rags.

2. Comic or graphic books: take apart and recycle with magazines because of the inks and paper.

3. Softcover or paperback books: roll and tie with string or cord. Use as kindling as long as the book does not have plastic.

4. Hardcover books: remove the covers and place them with your recycled cardboard. Rip up the book, roll it like a paperback, or put it with the recycled newspaper. Or use it as a craft project—I have seen authors make new books from the words on the page, and new stories and junk journals from older books.

5. Newspapers: roll them to make kindling or newspaper logs. Do not do this with magazines as the paper uses inks and sometimes plastics, which are toxic.

6. Magazines: recycle magazines, as they use chemicals in their print and their colored ink. You could also donate them to a doctor or dentist's office, or a school or library to use for crafts.

Maps

You can recycle maps with newspapers as long as they are not plastic-coated. If they are, recycle with magazines.

Photographs and photo albums

1. It is better to recycle photographs with your magazines as they have chemicals you do not want to release into the air. Shred photos of people.

2. Recycle photo albums like books—take off the covers and recycle them with cardboard and magazine pages. You should ask the family if they want pictures.

Music and videos

1. Commercial CDs and DVDs stamped in a factory last almost forever. They create great art projects, coasters and ornaments. Never dump, but donate wherever possible.

2. Homemade CDs or DVDs degrade on a microscopic and incremental level so that, over time, their readability decreases. One possibility is to hammer the homemade items into little pieces and use them as artwork or tiling. I have seen hanging CDs and DVDs without labels hanging from the ceiling. They reflect the lights or the sun.

3. Cassettes, CDs, and DVD jewel cases can be recycled with plastic.

4. VHS tapes should be donated if not recycled with plastic.

Tip D: Important Documents

No matter the paper or document, it should be shredded by hand or through a shredder. You can then recycle it with the newspaper or use it for packaging. Or, if worse comes to worst, you could burn it. We usually burn the mail we do not need. However, we only burn names, addresses, and account numbers, not the whole page.

The Third Principle: Sell

This principle, sell, was significant to Sasha and me. We made some nice money from Amazon and eBay listings on our media in 2006 and 2007. Selling your stuff on Amazon was relatively new back then, and you received some help from Amazon to sell books, CDs, and DVDs. eBay was the place for rare CDs, DVDs, and records.

Fast forward ten or more years. People still have CDs, DVDs, and books, but they are different. The postage prices have gone up, while the Amazon price for postage has stayed the same. More people have Amazon Prime and expect free shipping. Prices went down for music, videos, and books.

Not that you cannot sell media. You can, but you must be more diligent about prices, pictures, and descriptions. The things you sell must be perfect, with no rips, stains, scratches, or cracks. Let's get started!

Tip A: Take Pictures and Write Descriptions

Before you do any pictures or descriptions, look up similar items on Amazon and the sold listings on eBay. If they sell on either site for a penny up to $1, donate your items to a thrift store. Even at $1.01, you may not get your postage back on the item. It would be better to sell it at a garage sale. Sad but true—those $7.99 paperbacks are now worth less than the gum stuck under your chair. Selling popular books through the mail is not worth your time or money.

However, if you have a collectible book or other media, how you present them makes the difference. Use these tips to sell at the best price:

- Optimize the pictures with either white or black tablecloths as the background.

- Make sure you use directional lighting on your stuff.

- Prop up the object using boxes under the tablecloth.

- Write a description of your media and the shipping info, staying between twenty-five to two hundred and ninety characters.

Tip B: Figuring Out the Worth

While this may need to be done before taking pictures, most used books, CDs, records, DVDs, maps, and photo albums will cost close some where to the prices below. Media prices have gone down tremendously. Despite this, physical media retains demand, offering continued—though reduced—profit potential.

The average price of media:

- Maps: usually sell for a quarter—even the laminated and book ones.

- Photo albums: without the pictures, around $2.

- CDs, DVDs, and Blu-rays: common ones, between $1 to $2.

- VHS: pennies or are useless.

- Books: paperbacks, a dime to a quarter; hardcovers—fifty cents to a dollar.

- Rare, well-kept media: tens to hundreds of dollars.

Tip C: Other Places to Sell Your Stuff?

Photographs and scrapbooks

If the photographs are old and historical, you might get money for them in antique or other stores. You could also sell them on Etsy as a craft item. Look on Etsy for ideas on how to sell your photographs, scrapbooks, and maps. You still need to take excellent pictures and descriptions of your stuff.

Books

If your town has a used bookstore, you are a fortunate person! These are the best places to trade in your stuff. Many of these bookstores are in college towns, but they can be in other areas, too. I loved visiting the one in Moab, Utah; they have such an eclectic mix. Ask your local bookstore what their policy is for books and other media. Each store is different, but some will give you money, and others will give you credit for buying other books.

Maps and photographs

Historical societies in historical towns may pay for maps and photographs. Typically, they would rather have you donate them. You may ask for a fair price if the items are in perfect condition.

Garage sales, festivals, and flea markets are all places to sell your media in person. You must set up and pay vendor fees, usually for festivals and flea markets. But if you believe you will have more people looking and buying your stuff, go for it. This is also the easiest way to sell your stuff.

General tips for pricing ideas:

- Pricing can change from one year to the next. An artist dies, and everything sells at a higher price for a while.

- What stays constant are book prices at a quarter for a paperback, $1 for a hardcover, and magazines for a dime.

The Fourth Principle: Donate, Dump the Categories, Resources

Most of your media will be in this principle. People will not want your books, magazines, music, videos, photographs, photo albums, scrapbooks, or essential documents. And the donating rules have changed throughout the years.

Tip A: Donating Your Media

You probably have old-style media such as hardcover books, film photography or slides, VHS, cassette tapes, and DVDs. A couple of years ago, I entered a library where they covered the back wall with rows upon rows of VHS tapes and cassettes of audiobooks. The library allowed people to rent VHS machines.

Your VHS of the Titanic? Either find a retirement home with a VCR or accept that it's nothing more than a doorstop now. Or wrap the tape as a gag gift present.

Where to donate? Start with:

1. Libraries, schools, and nursing homes for videos, music, magazines, and books.

2. Historical societies, museums, national parks, and genealogy

studies will love your maps, photographs, and scrapbooks or photo albums, especially if you know the history of those items. The more knowledge you have about a particular item, the better.

3. Create your descriptions for each item with your name.

4. Get donation slips for your stuff and estimate the value.

5. Having a digital file of your donations, where they went, and their value is helpful during tax season and for posterity. You can also use the worksheet at the end of the chapter.

Tip B: Resources for Dumping or Discarding

Your community provides excellent resources for recycling and waste disposal information. Less stuff put into landfills is better for us all.

How To Organize Your Stuff

Once you decide how much physical media you want in your home, you need a place for each piece or category. Your display preferences may differ from your usage preferences. Make sure you can access every physical media type in an easy manner. An office or bedroom turned into an office allows for bookcases, filing cabinets, and desks with drawers.

Home without an office: Space is at a premium, so almost everything should be digitized. However, your prized books, maps, scrapbooks, or photo albums need a shelf and easy access. Store books in clear, shoe-box-sized containers under beds or in cabinets. Store DVDs and CDs in sleeve-binders and display the covers. Get an accordion-style, legal-size folder for your MOST important documents, i.e. marriage license, birth certificates, college transcripts, etc.

Home with an office: Use one side of the office for a floor-to-ceiling bookcase. Stagger the shelves to fit all of your physical media. Decide on enclosing the shelves with glass doors or not.

Invest in a desktop computer with enough storage to access your digital media and make backing up a priority. Have a small filing cabinet and safe combination for important documents. Get a desk with multiple drawers and allow yourself one as a junk drawer. All the other drawers should remain organized with your office supplies and physical media.

Garage space: Do not put most physical media in the garage, basement, or attic as most homes do not use temperature-control in those areas. Both digital and physical media can suffer. Reduce, digitalize, recycle.

How To Organize Your Physical Media

Here are some suggestions for the worksheets below.

Keep	Beloved signed books	Bookcase in Office
Keep	Important documents	Water-& fire-proof safe
Recycle	Wedding & Family photos	Digitalize!
Recycle	Tax documents	Digitalize!
Sell	Manuscript of German hymns	Christie's $8201
Donate	Maps of your town	Local historic society
Dump	Life magazines	Community craft or senior centers, or schools, but make sure to call ahead first.

You can find these tables by signing up for my email newsletter at LauraJevtich.com/OYSV2 and you can print these out over and over.

Organize Your Physical Media				
Keep		Recycle		
Item	Location	Item	Location	Reduce, Digitize, Recycle

Organize Your Physical Media			
Sell		Sell	
Item	Location	Item	Location

Organize Your Physical Media				
Donate		Dump		
Item	Location	Item	Location	Arts or Other

Hobbies

Applying the Four Principles to Your Activities

Hobbies are like that one kitchen drawer everyone has—it starts out organized and ends up a jumble of half-finished joy. My husband Sasha and I love hiking, writing, photography, videography, and playing water volleyball. Our friends love to sew and crochet, while others paint and write. I enjoy decorative planning and time management and use pretty-colored pens and stickers to create joyful plans.

What is the definition of a hobby? The dictionary describes it as "an activity done regularly in one's leisure time for pleasure."

Any way you slice it, hobbies are for pleasure. Once the creator earns a full-time income, the hobby becomes work. For example, an artist friend of mine makes her living creating art-to-wear clothing. She recycles metal bits to become jewelry and old t-shirts to become kimonos. She has a studio and a shop in California.

As a child, I had a collection of stickers to decorate my handmade cards and notes. As an adult, I've always had some hobby to keep me busy. In the big house, our basement became my artistry lair, with a Victorian dollhouse, paintbrushes, and glue in tidy shelves. Our tiny RV homes had storage for hiking poles and gear, and one cabinet for my decorative planners and stickers. In our casita home, I have an office with bookshelves displaying handmade cards, three desks with organized and colorful planners, and walls with pencil and watercolor artwork.

If your hobby is reading, listening to music, or watching videos, go back to the Physical Media chapter to see how to organize. If you are a gearhead or mechanic, go to the Tools for Home and Vehicles chapter. This section will be about sports, gardening, arts, and crafts. Get out your baskets, and let's begin.

The First Principle: Keep

Raise your hand if you've ever spent more time hunting gear than actually doing the hobby! Here's how to fix that.

Any physical activity where skill is required is a sport. This can also mean motorsports, ballet, and marching band. Yes, I lettered in the marching band in high school. If you have more than one hobby, find an area in your home to lay out all the equipment, materials, and pieces of work together.

Tip A: Similar Together

Organize similar items together by hobby. Figure out which hobbies you have and collect the items needed for that hobby. Lay the things out on a table or floor and go through them.

Tip B: Usefulness

Deciding how useful each hobby is will make this task easier. If you don't do the hobby anymore, it is easier to eliminate all the stuff within the hobby. Complete the process; provide frank feedback.

Have you used the equipment or participated in the hobby today?

↪ If not, have you used the equipment or participated in the hobby in the last week?

↪ If not, have you used the equipment or participated in the hobby in the past month?

↪ If not, have you used the equipment or participated in the hobby in the past six months?

↪ If not, have you used the equipment or participated in the hobby within the last year?

↪ If not, why do you still have the equipment or materials?

There's no reason to keep things you aren't using. You are just putting it off and being lazy, or you have no time to organize your hobbies.

Tip C: Containers and Storage

When you decide which hobbies you will continue, you need a storage place to house them. Some hobbies, like racecar driving and motocross, need garages. Other hobbies, like papercraft or writing, need shelves or drawers. Make sure you have enough room to store your equipment, materials, and finished products, whichever way you go.

Do not stuff items into a closet, where those items will fall on your head when you open the door. IKEA, Walmart, Target, Home Depot, Lowe's, and other stores have storage ideas for closets, cabinets, and drawers.

The Second Principle: Recycle, Rethink, Reduce, Repurpose

Have your hobbies changed over time? Does skill increase with time spent? Do you think you need other equipment or materials? This section requires a little more thinking or rethinking about your stuff.

Tip A: Rethink

Rethink your hobbies. Which ones bring you happiness? The definition of a hobby is that they provide you with pleasure. Ask yourself these questions:

1. If they are not bringing me joy, why have them?

2. If the hobby becomes too expensive and needs too much equipment, maybe it is time to give up the hobby.

3. Is the amount of equipment and supplies necessary for this hobby—and their storage, maintenance, and cleaning—overwhelming?

4. Do you still find you want to do this hobby, and can you do the hobby year-round?

Tip B: Reduce the Stuff

Are there ways to reduce your hobby materials? You might move from playing the piccolo to the flute, or recreational volleyball to league volleyball. Once your ability at painting becomes a mastery, your tools of the hobby grow more professional. In a sense, you are trading upwards. A local arts and craft store can buy back gently used supplies, or you can donate to schools and after-school programs.

Once you have put your discarded things into the different baskets, do not fill the space with more stuff! Enjoy the clean, open spaces.

Tip C: Repurpose the Stuff

Sport sneakers are great for workouts, but why not also use them for walking around? This is an example of repurposing your stuff. Your craft has extra fabric; why not use it as a cleaning cloth? The family's art brushes can help you with retouching paint on a corner molding. Your glue sticks can be used for repairing a picture frame.

Figure out additional ways to use your hobby equipment. Look into tools in the next chapter for those who have motor vehicles, woodworking, or maintenance as hobbies. Employ Pinterest to come up with new purposes for your stuff. My hiking poles became drying racks for clothes. Please be creative.

Examples

- A sport motorcycle (with a license plate) works as a weekend activity and as your commuter vehicle.
- Sneakers work well as walking and water aerobic shoes.
- Swimming suits work well as shorts.
- Photography mixed with photo editing software can become social media posts and art.
- Scrapbook paper in a planner adds extra depth and creativity, with pictures and quotes in a junk journal.

Tip D: Recycle

If the above tips do not work, recycle paper with newspapers or cardboard, and metal with aluminum or tin. Fabric can go with the Clothing, Shoes, and Accessories chapter. Plastic balls recycle well with plastic containers. You get the idea.

The Third Principle: Sell

Yes, it is time to let go of your hobby if it no longer sparks happiness when you play. Maybe you grew out of your hobby or moved to another place where you cannot do it anymore. All this happens for a reason. You can sell or donate everything you use in your hobby. Our tent, sleeping bags, and camping mattress sold well in Flagstaff, Arizona, right before a spring break one year. We had created our CamperVan, and no longer slept where the rattlesnakes lay.

Tip A: Take Pictures and Write Descriptions

With small items, take at least three pictures. For large items, take at least ten. Ensure the images are in optimal lighting, with a backdrop, and without other extraneous items. Pro tip: take photos of gear like you're listing it on a dating app—good lighting, no weird backgrounds, and describe it like the buyer knows nothing.

Tip B: Where to Sell?

If you have finished products for your arts or crafts, you could try selling them online. Etsy and Amazon allow you to create stores to sell your stuff and products. Zazzle or RedBubble do print-on-demand. You can sell the equipment to used sporting goods stores. Musical instruments to used music stores. Almost anything can be sold at garage sales if the price is right.

The Fourth Principle: Donate or Dump, Rethink, Recycle, Last Resort

Almost all items in this category qualify for donation. Find the people or the organizations who would want them.

Tip A: Donate

Donate whatever does not sell. Schools appreciate musical instruments and expensive sports equipment. If it has some small value but the time is not suitable for selling, it can be towed or driven away by nonprofit organizations. Starving artists and musicians treasure the materials needed for their arts and crafts. Most teachers, especially in the lower grades, need many craft supplies, and they would enjoy your items.

Tip B: Rethink

You might need to rethink or revise your ideas of throwing out a failed project. If you made something that destroyed your sanity, take it apart and reuse the pieces. Some famous painters have reused their canvases. You can rip out seams or unravel a stitch to reuse materials.

Tip C: Last Resort

I can't think of anything in this category that would need to be dumped, unless you do not have a way to recycle, incinerate, or donate. Make it as small as possible so the items can decompose faster, and bring it to the proper place.

How Long to Decompose?

- Paper towels: two to four weeks

- Drawing or crafting paper: two to six weeks in a compost bin, two to five YEARS in a landfill

- Non-waxed cartons (like egg cartons) for paints and small bits: five years

- Painted boards: thirteen years

- Lumber: ten to fifteen years

- Plywood: one to three years

- Leather: fifty years

- Metal (take it to a junkyard or dump to get the money):

    - Tin: fifty years

    - Aluminum: eighty to one hundred years

    - Aerosol bottles: two hundred to five hundred years

    - Iron: several years to a hundred years

    - Aluminum foil: never

- Styrofoam: never

How To Organize Your Stuff

This is a vast category, with everything from motor vehicles to gardening, physical activities to stationary activities to arts and crafts and more. Therefore, generalizing how to organize your hobbies requires thinking outside the box. Use suggestions from other categories to help organize.

Hobbies requiring indoor space: Stationary and arts and crafts hobbies need tables or desks, comfortable chairs and lighting; maximize the amount of flat surfaces with desks and drawers. Keep the supplies in dedicated drawers to minimize stuff on the flat surfaces.

Indoor physical activities need the space to do said activity. Use the garage or basement with no clutter nearby. Invest in foam interlocking squares for exercise on concrete or hardwood surfaces.

Hobbies done outdoors that need indoor storage space: Store sports equipment in the basement or garage, using pulley systems and shelving for large equipment. Keep equipment off the ground in case of flooding. Use corkboard or pegboards on the walls.

Extraneous spaces: Sports or activities done outside the home may have lockers or trailers for your equipment, which you can rent. Do this as needed.

How To Organize Your Hobbies

Here are some suggestions for the worksheets below.

Keep	One hobby keeping you sane	Somewhere close
Recycle	Sick of hobby (know friends) (don't know anyone)	Reduce or Put into a large Free box at your next garage sale
Sell	Kayak (doesn't fit bike)	Buy, sell, trade $250
Sell	Plastic planters or pots	Give away with each plant bought at a farmer's market
Donate	String Bass	Music Dept at local school or college
Donate	Pencils, markers, pens	Schools or Free Box
Dump	Hemp sketch paper	Community or home compost bin (cut paper into tiny pieces first, then wet pieces before dropping paper in.)

You can find these tables by signing up for my email newsletter at LauraJevtich.com/OYSV2 and you can print these out over and over.

HOBBIES

Organize Your Hobbies				
Keep		Recycle		
Item	Location	Item	Location	Refresh, Repurpose, Reuse

Organize Your Hobbies			
Sell		Sell	
Item	Location	Item	Location

Organize Your Hobbies				
Donate		Dump		
Item	Location	Item	Location	Arts, Music or other

Tools for Home and Vehicles

Applying the Four Principles to Metal Doodads

Let's be real: tools are the unsung heroes of adulthood—until you're staring at seventeen identical wrenches wondering which one "sparks joy."

Each time we moved, Sasha and I argued over how many tools we took with us. It seemed to me he was taking them for no reason. How many wrenches do you need? However, when I realized he knew where the tire jack was when we had a flat tire, I stopped questioning his need for tools.

Tools make our lives easier and more livable. I sometimes wish there was one tool to do all the jobs.

When going through your tools, I suggest you discuss and explain to your partner why each holds value. If you have multiple versions of the same tool, have trouble finding them when needed, or haven't used them for over ten years, it might be wise to get them out of your life.

The First Principle: Keep

Retrieve your tools and consolidate them using the baskets. I would suggest doing this either in a garage, a basement, or outside. Realize when you putter around in the garage, more people will come by with questions and discussions. Your organizing might get lost with tool talk.

Tip A: Similar Together

Put similar items together: all the hammers, wrenches, nails, screws, sockets, screwdrivers, nuts, bolts, saws, etc. When doing this, look them over to ensure no rust, duplicates, or missing pieces. Test the mechanical or electrical ones to make sure they work. Emotions need to be checked at the door. That machete from Uncle Joe's "jungle phase"? Unless you're secretly prepping for the apocalypse, let it go.

Tip B: Usefulness

This tip is for all those who hang on to those tools. I know you might need it somewhere down the road. Your uncle or your mother gave you this or that tool because they also owned one and thought you should have one.

Ask yourself these questions:

Have you used this item in the past week?

↪ If not, have you used it in the past month?

↪ If not, have you used it in the past year?

↪ If not, have you used it in the past decade?

↪ If not, have you EVER used it?

If you answered no to all these questions, why do you still have it? Yes, you may have forgotten you had it, or maybe you received it from some well-meaning person. Your guilty emotions tie you to that tool. Tear yourself away from the feelings and be ruthless in the devices you use.

Tip C: Containers and Storage

Thankfully, many tools have their handy-dandy kits and containers. Or you can get toolboxes to store everything. My father has a workbench in his garage. This workbench is beautiful and practical, with a cork board where you could hang tools. The most important part was the labeling. Each tool had to be returned to its spot on the corkboard, making it clear if anything was missing. Having a place for everything and everything in its place links to a happy home.

The Second Principle: Recycle, Repair, Reduce or Share, Rent

Here comes the moment of truth. Some of your tools are broken or rusted, or they're missing pieces, or they've never even been used. You have the nuts and bolts for a motorcycle you had twenty years ago, and the socket wrench to go with it. Recycling your tools can be easy, especially if you have quality tools. Many companies have generous return policies. If a tool is rusty, do not recycle or sell it. Go to the Fourth Principle.

Tip A: Repair

If your tools do not turn on, or if they only work intermittently, repair may be the way to go. If you've registered your device, you might be able to get free or cheap parts for that tool. There are forums and user websites and videos on YouTube where you can learn how to repair your tool. If a community college or vocational school is nearby, you might try going there to fix your device.

Tip B: Reduce/Share

Pro tip: make friends with your neighbors. When we first moved into the big house, we shared tools like a communal candy bowl—until Dave "forgot" to return our drill. Lesson learned.

If you live in a small community, sharing is common, which might work well for you. We know a gentleman who carries welding equipment in his rack on the side of his RV. He is generous with his time and capabilities. Sharing saves you money in the long run and leads to community.

Tip C: Rent

For those who don't have the space or only need the tool for a short time, renting is an option. I mean, who owns a carpet cleaner, anyway? You can rent by the hour or day. Sometimes going to or calling the rental shop is beneficial, even if you don't know the tool's name. Describe the task you want to accomplish, and they can suggest the tool you need.

Example
When we had the big house, we rented a pressure washer to remove the stains from our deck. Our sisters rented generators when their power went out. Rentals can be successful space savers; plus, you can try out the tool before you decide to get your own.

The Third Principle: Sell

While most of the stuff you sell on the secondhand market does not keep its value, tools can be the exception. Tools do not sell for retail prices, however, tools sell on the secondary market and keep their value. We found many tools at the estate sales used throughout a family's life. You can find deals in the secondary market and sell your tools.

Tip A: Take Pictures and Write Descriptions

Clean the tool as best as possible before you take your picture. Take pictures on a white or black background. If it is a power tool, take a video or a series of images with it on and working. Describe the device and how you used it.

Tip B: What is it Worth?

Pricing tools is like dating—you need to know the market. Check garage sales and eBay, but when in doubt, ask that one DIY-obsessed uncle we all have. When I was younger, my uncle and dad would be in the garage or the basement during their estate sales, pricing the tools.

Many tools sell on Craigslist and eBay, but the best way to find prices is by checking out garage and auction sales in your area. Going to flea markets and swap meets would also be beneficial as tools prevail in those avenues. Each type of tool will be different depending on user care.

Tip C: Where to Sell?

Where do you sell? Well, it's the same place you researched prices. Try a garage sale before selling it at flea markets, because flea markets have vendor fees. If you have a lot of tools, especially of the power kind, it is best to do so at a garage sale. Make sure to have extension cords plugged into outlets to show customers that the tools work. Otherwise, you would have to bring a battery to show how they work.

The Fourth Principle: Donate or Dump, NO!, How Long to Decompose, Repurpose

Please do not dump anything in this category.

Tip A: Donate

- Local thrift stores (especially for power tools)
- Community colleges
- Gardening clubs, High school shop classes (future builders thank you!)
- Artist collectives (metal = free sculpture supplies)
- That neighbor who always borrows stuff (tag it "LOANER" and forgetaboutit)

You could also gift the tool to your gardening club or family, especially when it is a tried-and-true tool. This might become a tradition, like it did in our family.

Tip B: NO!

Confession: I'm a tool recycling fanatic. That broken drill? Its parts could live on as edgy earrings. (No judgment if you're not that committed.) Or you could cannibalize the parts to repair other tools. Most tools, particularly metal ones, are recyclable. Make sure to take off any plastic pieces and recycle them with your hard plastics.

Tip C: How Long to Decompose?

Metal and plastic tools take anywhere from a hundred years to never. It takes more energy to remove them than to make them. That being said, rusty metal tools brought to scrap metal junkyards can earn money.

What happens to metal components in vehicles and tools? The plastic in a tool or vehicle gets removed and recycled. The metal in a tool or vehicle becomes shredded, melted, or crushed. Metals are disposed of the same way they come into the world: smelting and burning at high temperatures. Recycling areas today have more control and regulations than in the past. These processes can provide energy to the surrounding community.

Tip D: Repurpose

If the tool breaks beyond repair, some artists take the pieces and make large art displays. Metals such as aluminum, copper, and brass can be used to make jewelry. You can also give metal scraps to schools, high schools, and colleges to be used for arts and crafts—they would love the free materials.

Example
However you purge, remember: the goal isn't a Pinterest-perfect garage. It's about making room for the tools you'll actually use—and maybe, just maybe, finding that missing socket wrench. Remember, too, that not everyone can be my dad, with his perfect workbench. Most are like my late uncle, who had a cluttered workstation but still knew where every tool was in his collection. However, he did not know if someone borrowed a tool or tools and when he would get them back. Which is why he did not lend out his tools very often.

How To Organize Your Stuff

Put tools in easy-to-access areas. Put vehicle tools in the garage and household tools in the laundry area or kitchen, and employ wall space as much as possible.

All tools, whether for the home, landscape, garden, or vehicle, need to be maintained and stored close to whatever it is they're used for. Labeling—use a label maker for easy-to-read labels—tool boxes, and wall storage allow you and your family to put the tools back in the right place. Create maps or sketches of storage areas to make sure it all fits.

How To Organize Your Tools

Here are some suggestions for the worksheets below.

Keep	The standing toolbox	Garage back wall
Keep	The set you always use	Laundry or Bath
Recycle	Trenching machine	Rent
Sell	Mysterious screws in drawer	Craigslist $5
Donate	Handy tools	Graduate student
Dump	Rusty tools	Scrapyard $10 for the metal

Most tools can be kept for Years and Decades. But if you no longer have a motorcycle or car, give the new owners those tools to fix the machine. You'll have friends for life.

You can find these tables by signing up for my email newsletter at LauraJevtich.com/OYSV2 and you can print these out over and over.

TOOLS FOR HOME AND VEHICLES

Organize Your Tools				
Keep		Recycle		
Item	Location	Item	Location	Repair, Reduce / Share, Rent

Organize Your Tools			
Sell		Sell	
Item	Location	Item	Location

TOOLS FOR HOME AND VEHICLES

Organize Your Tools				
Donate		Dump		
Item	Location	Item	Location	Price

Electronics

Applying the Four Principles to Your Digital Life

This chapter covers items that run on electricity, (excluding items already covered in the household goods chapter) including TVs, computers/laptops, phones, stereo systems, electronic systems, wires, Kindles, game consoles, cameras, video equipment, and more.

Example

We had a vast collection of electronics when we lived in the big house. We had a state-of-the-art electronic system. Our stereo system and video/film equipment rivaled those of a small company. Our home network had over a terabyte of storage.

We were among the first to install a cut-out over the fireplace for our Samsung DLP rear-projector TV. Since Sasha worked in the audio-visual industry, he knew the latest and greatest electronic equipment and got deals. It's a smart idea to get deals since electronics, no matter the quality, are worth about ten cents to the dollar in resale.

I realize it will not be easy to group similar items together, especially with large items like TVs or computers, but let us try so that you can keep what you need.

The First Principle: Keep

Our Examples

Even though we had an extensive electronic system in our big house, we were sparing with similar items. We owned one TV, and that was in the family room. The house had thirteen to fourteen speakers connected to one stereo system that synced up perfectly in every room.

While we did not have as many electronics in our tiny RVs, we still had speakers throughout, with one TV and one stereo system.

This mantra of minimalism works for our small casita: one smart TV, one stereo system, four speakers in the main room, and speakers outside. I decided not to have speakers in the bedroom or craft room. Sasha has a music system set up in the big garage, as well as a mini-computer and monitor that we use to play exercise videos.

You must decide what to keep, recycle, sell, donate, or dump. Gather your baskets and let's get started.

Tip A: Similar Together

First, go to the room in your home that has the most electronics. Look for anything that uses electricity that is not a kitchen or bedroom item. This includes:

- TVs (including smart and analog tube TVs)
- Desktop computers and laptops
- Kindles and e-readers
- iPads and tablets

- iPods and music players
- Smartphones and regular phones
- Calculators
- Printers and scanners
- Answering machines
- Radios
- Stereo components such as CD players and subwoofers
- DVD and Blu-Ray players
- Digital video recorders and speakers
- External hard drives
- NAS (Network Attached Storage) and their components
- Cables and wires
- Power supplies
- Game consoles
- Streaming devices
- Components and external equipment
- Cameras and their equipment
- Audio (electronics such as headphones, speakers, and microphones)

ELECTRONICS

- Lamps or lighting used for creating videos or photography

Put similar items together. Don't forget to look at all the components of each item, such as the power cord and battery pack that go with your laptop.

You have decisions to make here. Prioritize your current needs. Will your decisions maximize your enjoyment, savings, and free time? Or will holding on to these items make you crazy, withhold your happiness, and cost you money?

How old are some of these items? I was going to add fax and adding-machines—I've seen them in doctor and dentist offices, and NOT that long ago—as well as answering machines and landline phones. But I took them out, realizing most readers of this book got rid of them long ago. You can often be ruthless and without emotion in deciding what to keep in this category. With that said, let's figure out the usefulness of these items.

Tip B: Usefulness

This is where we will cull the useful from the "WOW, I did not know I had that item," and the "SO THAT'S where it is!"

Ask yourself these questions:

Have you used this item in the past day?

↪ If not, have you used this item in the past week?

↪ If not, have you used this item in the past month?

↪ If not, have you used this item in the past six months?

If not, why do you still have it?

Electronic items update every six months. They are still usable after that time. But if you had them in a drawer or closet somewhere, more than likely you could get rid of them instead.

Tip C: Containers and Storage

Many of the items discussed in this chapter do not move from room to room. Your TV or DVD player stay in your entertainment center. Your cables and wires stay behind the cabinet, and you never see them again, at least until you add a new item to your stereo system or upgrade your TV. Then you need to move furniture and confront the mess of wire, battery packs, and cables going to who knows where. Now's the moment to solve them.

Use cable ties, cord splits, and outlet strips to organize your equipment. When setting up our big home network, Sasha labeled the wires so we would know which ones belonged to each stereo or computer network.

If you have a laptop or desktop computer, store the parts near them. We usually store them in backpacks, covers, and storage bags. Covers for movable electronics help secure your stuff. Cases for hard drives, smart phones, laptops, etc. keep them secure and clean and put the cables close to the equipment. The best place to buy essential covers or cases is Amazon. If you want to have your designs or artwork, use Zazzle or Red Bubble for that. You can get split cables and outlet strips at Walmart, Amazon, or Target.

The Second Principle: Recycle, Reduce, Reuse, Repair

Recycling your electronics, whichever way you do so, is always better than throwing them away. The number of chemicals leaching out of discarded electronics could fill a sea, and it's only getting worse. Here are three tips to recycle your electronics.

Tip A: Reduce

You can start preventing and minimizing waste before you even bring an item into your home by purchasing high-quality, durable, and upgradable electronics. By doing this, you save money and reduce waste in the long run.

Example
Sasha and I have bought refurbished laptops from Apple. The second laptop we owned was a refurbished 2012 MacBook Pro, which was the last model Apple sold that still had upgradeable RAM, hard drives, and batteries. Unfortunately, Apple decided to go slim and no longer makes durable and upgradeable computers.

Reducing waste with electronics can be as simple as turning off monitors, stereos, or TVs when unused. Turn off your printers, copiers, smart phones, and laptops until you need to use them.

Tip B: Reuse/Repurpose

Reusing and repurposing can work together. You can reuse your printer or other electronics by changing the ink cartridges, battery, or screen. Repurposing is a little different; you use the item or your stuff differently. We have a Mac Mini, a mini-computer that you can add a monitor, keyboard, and mouse to in order to make it work like a desktop computer. We repurposed it to play back our videos and music. It is the connection between our controls. We can also use it to listen to music.

Most electronics do not share batteries. However, some do, allowing you to repurpose your cables for other electronics. Check the ends of each cable to see if it could be used for something else. I've used cables to charge headsets, cameras, and e-readers.

This tip would also include using or renting from others for some of your printing or scanning needs. Check your local print shop to see if you can get your large-scale printing done there.

Tip C: Repair

If people did not repair their expensive electronics, our landfills would already be at capacity. Some people have taken this to a high level. The Geek Squad and Apple Genius Bar help you use your computers and their components for longer. Repairing a $1,000 electronic is a much better idea than buying a brand new $1,500 replacement.

If you cannot fix it yourself, put it in the basket for recycling to get it repaired. You might also want to check out the "Right To Repair" legislation in the U.S. The link is.

Repair can also mean upgrading. Our iPod, which we use for our music storage and listening, had a problem with freezing up. We sent it out to a shop that reformatted and upgraded the equipment. It costs less, and it gives new life to our almost-obsolete electronic equipment.

The Third Principle: Sell

I will tell you right now—the high-end electronics we had in our big house NEVER paid us back when we sold them. We enjoyed our use of the items. But when we sold them later, we received a small return on the initial investment.

The only things that did sell for the same as we bought them for (and we believe they were a couple of flukes) were our iPhone 4 smartphones and two MacBook Pro laptops. It might have been because we sold them in a college town, or because it had been a while between the updates. I'm not sure. We sold them via Craigslist in the resale market.

Tip A: Take Pictures and Write Descriptions

We took excellent, detailed photographs and wrote flawless descriptions for Craigslist. The item was shown both turned on and turned off, which is important for you to be able to sell it at a higher price. You'll also need to state in the description if you're willing to sell it for the parts. We did that with a laptop that wouldn't turn on, and we got a reasonable price for the parts.

Tip B: What is it Worth?

This can be heartbreaking, especially for those who upgrade every six months to a year. While you never get your investment back, if you buy it and then try to resell it six months later, you might get a price close to retail. But as soon as those six months are gone, the worth decreases at a rapid rate, almost exponentially. Look on Craigslist, eBay, and Amazon to figure out prices. Do not use your retail price as the starting point.

Tip C: Where to Sell?

The best place to sell is usually Craigslist. Craigslist does not eat away at your profits, either.

Hands down, this or your Facebook Marketplace or community resell boards are best. The second best would be Amazon, eBay, or Mercari, but you must deal with shipping and commissions. Sometimes, having a smart

TV on playing music while having a garage sale will get people to buy more. This may encourage them to make an offer on the TV.

The Fourth Principle: Donate, Don't Dump, Wipe Clean, Rethink, Repurpose or Scrap Yard

Tip A: Donate

Please, please, please, do not donate your analog tube TVs or old electronics to schools, nonprofits, or thrift stores! That is mean. These places must then determine where to recycle.

Give your electronics to schools, nursing homes, or thrift stores as long as your electronics are not old. Electronics under three years old would fit this. Anything over three years old would require more of an explanation on how to use it for the recipients.

Example
We had donations of useless electronics at the nonprofit we worked for, and it became a huge hassle. While the donor got a write-off, we did not make any money, and we spent money to get rid of it.

Tip B: Don't Dump

Throwing out electronics is a serious NO-NO in our household. As a trained environmental geologist, my job involved cleaning up hazardous waste, heavy metals, chemicals, and petroleum products.

Heavy metals such as cadmium, chromium, mercury, and polychlorinated biphenyls (PCBs) were all in my realm of hazardous waste management. They were nasty in the soil and even nastier in the groundwater. Throwing away electronics into landfills because you are lazy hurts everyone else. So please take my advice and don't do it. Future generations and your children and grandchildren will thank you.

Tip C: Wipe Clean

Most electronic devices will have your information on them. Do you want the new owners to know your passwords or to have access to pictures of your family? Wiping your personal identification from your electronics is a bright idea. Also, you can sell them for more money, especially if your phone is "unlocked." Unlocked means capable of operating on any network or service provider. Other options are to reformat cameras, hard drives, and other data storage methods.

Tip D: Rethink Personal Printers

An eco tank printer instead of a disposable ink cartridge printer goes so much further in the long run. You do not have to dump those ink cartridges as you refill the tanks. And they work well for home use.

Tip E: Repurpose or Scrap Yard

As the tools chapter states, you can take your electronics to many scrap metal yards. They break apart the pieces, separating the plastics from the metals. They may give you money for the copper and other metals or charge you for bringing your electronics. Some towns have an electronics recycling day where you can bring your electronics in. They will give you a receipt to use for tax writeoffs, and you'll be helping protect your town and its landfills from hazardous waste. Doesn't that feel good?

How To Organize Your Stuff

Electronics may be all around your home or in your vehicle. Keep electronics and related containers in drawers or cabinets. Store the accessories for each electronic piece in clear bags or strong, cotton containers. Create a room-by-room inventory of electronics. Use the worksheet below.

Homes with one central smart TV: a smart TV means one with the capacity to "speak" to your computer, connect to the internet, and integrate features with one or two remotes. Sometimes placed on a cabinet with drawers or shelves, smart TVs can also be hung on the wall. Audio equipment integrates with the TV, allowing surround sound through speakers and subwoofers. By placing these two major electronic systems together, there is more space available for other furniture.

Homes with multiple TVs, smart or not: as soon as you add more TVs and their accompanied components, you need more furniture to store the extras and the accessories. Make sure the internet coming into the home is powerful enough for multiple Wi-Fi electronics, otherwise your household will complain or get frustrated. Label the wires coming and going from the electronics to other components. Map or sketch each room with electronics and their wires. Otherwise, you will have a wire mess to sort through when you take away or add a new component.

Computers or other electronics with operating systems: when you have more than one operating system, Windows, macOS, Android, Linux, or iOS, maintain these devices with weekly clean-ups, dusting, and back-ups. Buy covers and sleeves to protect the devices while traveling. Keep them charged with a full battery, just in case you need them.

ELECTRONICS

How To Organize Your Electronics

Here are some suggestions for the worksheets below.

Keep	Latest & Greatest iPhone	In office or hand
Keep	Chargers for owned electronic	Drawer
Recycle	Child's laptop	Repair & upgrade
	Chargers from sold electronics	Most communities have a recycling day for electronics, or Best Buy, Office Depot, Staples, Apple, Freecycle. Others can include some Walmarts.

Sell	Multiple smart TVs	Garage Sale $100
Donate	Tablets	Local school or library
	Older e-readers	Senior Centers
Dump	Unable to turn-on electronics	Geek Squad or your local electronics repair shop (they work with recyclers for electronics.)

	Organize Your Electronics			
Keep		Recycle		
Item	Location	Item	Location	Reduce, Reuse, Repair

ELECTRONICS

Organize Your Electronics			
Sell		Sell	
Item	Location	Item	Location

Organize Your Electronics				
Donate		Dump		
Item	Location	Item	Location	Price

ELECTRONICS 157

You can find these tables by signing up for my email newsletter at LauraJevtich.com/OYSV2 and you can print these out over and over.

Food and Provisions

Applying the Four Principles to Your Supplies

This category discusses food, paper, cleaning products, liquids, and water, and it's a chapter near and dear to my heart. In all the homes I've lived in, I made sure I knew what provisions and supplies were in the home. Having two documents made this easier: a map or sketch of our food and provisions, and the worksheet below. This has allowed us to spend more leisure time doing fun activities and adventures rather than shopping for food and short-term items. We rotate the food we eat and store and do not have containers of green goo or scientific experiments. No mold or strange, colorful spots on our food.

Example
Both Sasha and I came from families who made enough food at each meal for plenty of leftovers. There is nothing wrong with leftovers as long as you organize and manage them. I think the tradition of keeping the top layer of the wedding cake ended when someone forgot about it in the freezer. She pulled the frost-covered cake out of the freezer, hoping to eat it and reminisce with her beloved about their wedding, and instead they got food poisoning.

My aunt and uncle once had a freezer filled with leftovers. While they dated the food, I believe one lasagna was in there for about twenty years!

Gardeners know their canning and storage capabilities. The most important lesson I learned from growing up with gardeners is stock rotation. Use the oldest items in your pantry, refrigerator, freezer, basement, closets, and other areas first. Writing dates on the top of the containers or labels makes this process and rotation easier.

A cool day is ideal for checking your food supplies. Make sure you have your baskets ready.

The First Principle: Keep

Start with one area of your food storage, such as the refrigerator or freezer, and then go to cabinets, pantry, basements, ancillary refrigerators and freezers, and other places where you have foodstuff, paper, and cleaning supplies. Go to one area at a time to decide whether to keep, recycle, sell, donate, or dump.

Say you start with the refrigerator. Get removable labels, a permanent marker, and a pen or pencil. Take everything off the shelves, bins, and doors.

Tip A: Similar Together

When you sort, put like items together—forming groups of fruits, vegetables, dairy, meat, grains, condiments, and liquids—and inspect the food as you do so. Is mold growing inside the containers? Do you even know what's in there? Put all the questionable items together. We will not dump them yet. Get coolers for the meat and dairy, plus the prepared foods.

Once all items are in similar groups, get your permanent pen and labels. If you can, figure out when you bought the item and put the date on the label. Reduce text size; this allows multiple future dates on the tape. Label. Before you put the items back, decide upon their usefulness.

Tip B: Usefulness

Are those pickles from a year ago, or are they from last week? Now that you have the dates, it is easier to remember when you last used and or ate it.

Ask yourself these questions:

Did you eat or use this food during your last meal?

↪ If not, have you eaten or used it in the last day?

↪ If not, have you eaten or used it in the last week?

↪ If not, have you eaten or used in the last two weeks?

↪ If not, have you eaten or used it in the last month?

↪ If not, is it meant for long-term storage?

Do you have to store it in the refrigerator or store it in a cool, dry place? Do you need it?

Figure out what is still useful to you and organize the refrigerator. Prioritize older items; place newer ones behind them.

Tip C: Containers and Storage

When you buy foods in bulk and separate them into different containers, you need to decide what type of containers you need. The lightweight plasticware we used in the RV came in many shapes. The plastic was BPA-free. Make sure when you buy containers that they are safe for food storage.

There are some heavy-duty food storage containers where you have to hammer down the top to close them. These are for long-term storage of bulk foods, usually separated by zippered plastic bags. Typically, we might have one bag of black beans and another bag of a different bean, but they are all in one container. A label or date on the bag helps us figure out which needs to be used first.

For bulk storage containers, we bought from grocery stores that sell bulk-sized foods. You can find them at Aldi, WinCo, Costco, and Walmart. Rubbermaid is a specialized brand of food and liquid storage. I would suggest checking Amazon and other online food storage stores.

When you buy items from big-box stores like Costco and Sam's, they may have containers that are great for long-term storage. Be aware of old plastic containers. These older containers produce off-tasting toxic chemicals, especially over time. One of the things we use most frequently are water bottles., Say goodbye to toxic, plastic water containers and use stainless steel containers instead.

The Second Principle: Recycle, Reduce, Refuse/Reject, Reuse

Remember how I said you shouldn't put the green goo and other experiments back in the refrigerator? We are going to try an experiment of said goo. Let's see if we can reduce waste first.

Tip A: Reduce Waste

Before you get rid of that green goo, are there items in your cabinet or refrigerator that you bought for a recipe but never used? I have done this before. Next time you go grocery shopping, bring the worksheet at the end of the chapter. That way you'll know what you need, and you'll stop waste before you even begin.

Reducing waste means you need to figure out your eating habits as a family. Do you eat more fruits and vegetables during the spring and summer? Can you freeze, can, or dry extra food before it rots? We used a lot of dried food in our tiny homes, as it weighs less and can be stored for a longer period. We continued the practice in our small casita home.

You can reduce the waste of foods by buying in bulk and choosing foods that last longer. These foods and drinks either have a longer shelf life or need to be refrigerated once opened. We also have an auxiliary refrigerator, which helps store food longer.

Tip B: Refuse or Reject Packaging

We find and buy less at the big-box stores because of all the packaging. A ten-pound box of oatmeal might come in a cardboard container with four plastic or wax bags. Why not buy ten pounds in bulk? I got sick and tired of breaking down all that cardboard and finding a place to recycle it. I also dislike how they give you cardboard boxes to put in your vehicle. They give you their trash so they'll have less to recycle and break down. Refusing their boxes and bringing recycled bags (maybe even bags you made yourself) allows you to bring less home.

Tip C: Reuse

Buy large boxes and larger sizes of cleaners, cooking oils, vinegar, and other liquids. You can then put them in smaller containers later. I reuse the smaller containers. They are easier to handle with fewer spills and waste.

The Third Principle: Sell

Selling your leftovers is not something I would recommend, unless you start a side business by finding people who like your cooking. Even trading or bartering your foodstuff can yield benefits. You need pictures, descriptions, knowledge of your county laws, and places to sell.

Tip A: Take Pictures and Write Descriptions

Foodstuff looks tasty on a white tablecloth. Use either natural light or designated lights. Take the photographs without a flash. Include pictures of the ingredient list and the calorie count with the nutritional information.

Tip B: What is it Worth?

You may get acceptable prices if your baked goods, dried seasonings, or three-bean salad mixtures are winning at the county fairs. Figure out the prices you paid for individual ingredients on sale at the store. Take a ride in the country and check out roadside stands. Look at their prices and divide by your price. Why? Because they have a store and overhead for their price.

Tip C: Where and How?

You need a license in your state and sometimes your county when selling food. Alternatively, you can use a certified kitchen. You might be better with baking or uncooked bean mixtures or products. If you get a vendor's license or food handler's permit, think about following the requirements.

That said, there are places you can go to sell your foodstuff, such as festivals, fairs, community or church events, and others where you can sell your food. Sometimes it's easier to sell baked food or foods in jars like jams, jellies, and vinegar. Those items are considered "products" rather than regular cooked food.

Start checking Craigslist, online news, or Facebook groups and pages in your area. Set up a booth to showcase your food, and see how you do. When the profit exceeds your expenses, you may have found a hobby.

The Fourth Principle: Donate, Dump, Last Resort, Rethink, Statistics, How Long to Decompose, Repurpose, Resources

Tip A: Donate

Uncooked food can go to shelters, food pantries, and other organizations. Bake sales for veterans, churches, and schools are considered donations because you give your time and ingredients, and the finished product and the proceeds go to the organization. Take pictures and write descriptions, including the calorie count, so that it's easier to sell.

Are you a member of the Clean Plate Club? Parents try their patience to get you to eat green vegetables, and they tell you of the starving children in XYZ. My sister would say, "Give my vegetables to them." However, many other parents allowed those vegetables to be dumped into the trash.

Tip B: Last Resort, Rethink, and Statistics

Food waste has reached high levels. In the U.S. alone, sixty million tons of food waste land in landfills each year according to 2025 records. Living in suburbia and cities makes dumping food so much easier. But if you belong to a food co-op, or if there are community gardeners around, you might see if you can bring your food waste to them.

The best way to not throw out food is to not make enough for leftovers. Determine what your family eats for any meal and give them that portion on the plate. Make the plate smaller and tease the mind into thinking there is enough on the plate.

Play around with different routines, food types, and quantities until you figure out what you need. Buy only what your family eats before the food spoils. Eat foods that store well. These foods include fruits, beans, dried milk, powdered juices and drinks, fresh apples, oranges, cabbage, onions, carrots, and potatoes. Canned meat such as tuna, chicken, and beef store well. Once opened or hydrated, they need to be refrigerated. Remember to rotate old to new.

Tip C: How Long to Decompose?

How long before food decomposes in a landfill?

- Vegetables: five days to one month
- Bread: two to three weeks
- Paper towels: two to four weeks
- Paper waste: two to six weeks
- Food waste: several months to several years

- Apple cores: two months

- Banana peels: up to six months

- Orange peels: six months

- Waxed milk cartons: three months

- Eggshells: one to ten years

- Nutshells: one month to three years

- Non-waxed cartons: five years

Tip D: Repurpose

A brief lesson on composting and vermiculture: both decrease food waste by creating energy and enriching soils and plants.

Vermiculture or vermicomposting is the use of worms to process organic food waste into nutrient-rich soil, liquid, and castings.

Composting is the natural process of recycling organic waste. Organic waste includes brown material to produce carbon (dead leaves and twigs, coffee filters, cotton and wool rags, shredded pieces of paper, cardboard, newspaper, sawdust* or wood chips*) and green material to produce nitrogen (grass clippings, green leaves, fruit and vegetables, hair, lint, tea and coffee grounds) and water plus aeration. Do not put cooking oils, pet waste, fried foods, meat scraps, or eggs into the pile.

There are two strategies to composting—active, which is hot or temperature-managed, and passive, which is cold or static piles. Active composting uses heat, compression, and organic matter or food waste to make nutrient-rich soil. When you cut the waste into small pieces without liquifying it, microbes go to work and break down the ingredients.

Both methods work well for gardens and potted plants. Organic compost—the rich-brown to black material—may sell well in your area. The knowledge you receive is almost worth the hassle of trying this out.

Tip E: Resources

Find your local master gardener extension office at mastergardener.extension.org/contact-us/find-a-program and email or call to find offices in your town. Places such as KingArthurBaking.com and Lehman's make products to store food. Other places include: Walmart, Amazon, The Container Store, IKEA, and your local hardware or gardening store.

How To Organize Your Stuff

Refrigerators, pantries, and cabinets in the kitchen are the most common places to store your food and provisions. See below for worksheets: one for the process of reducing the amount of stuff. The other sheet is graph paper you can use to map out where the items are within your storage areas. Then sketch the rooms where you stored the foodstuff, especially if in unusual places.

How To Organize Your Food & Provisions

Here are some suggestions for the worksheets below.

Keep	Your mom's lasagna	Kitchen freezer (dated)
Recycle	Today's chicken	Reduce waste & make Chicken salad & use bones for soup
Sell	Dried excess fruit	Local farmer's market
Donate	Excess garden veggies	Friends or food bank
Dump	Extra fruit scraps	Compost or give to worms

* Do not use pressure-treated wood, as the process is toxic.

Organize Your Food & Provisions				
Keep		Recycle		
Item	Location	Item	Location	Reduce, Refuse, Reuse

ORGANIZE YOUR STUFF

Organize Your Food & Provisions			
Sell		Sell	
Item	Location	Item	Location

FOOD AND PROVISIONS

Organize Your Food & Provisions				
Donate		Dump		
Item	Location	Item	Location	When?

Map of _____

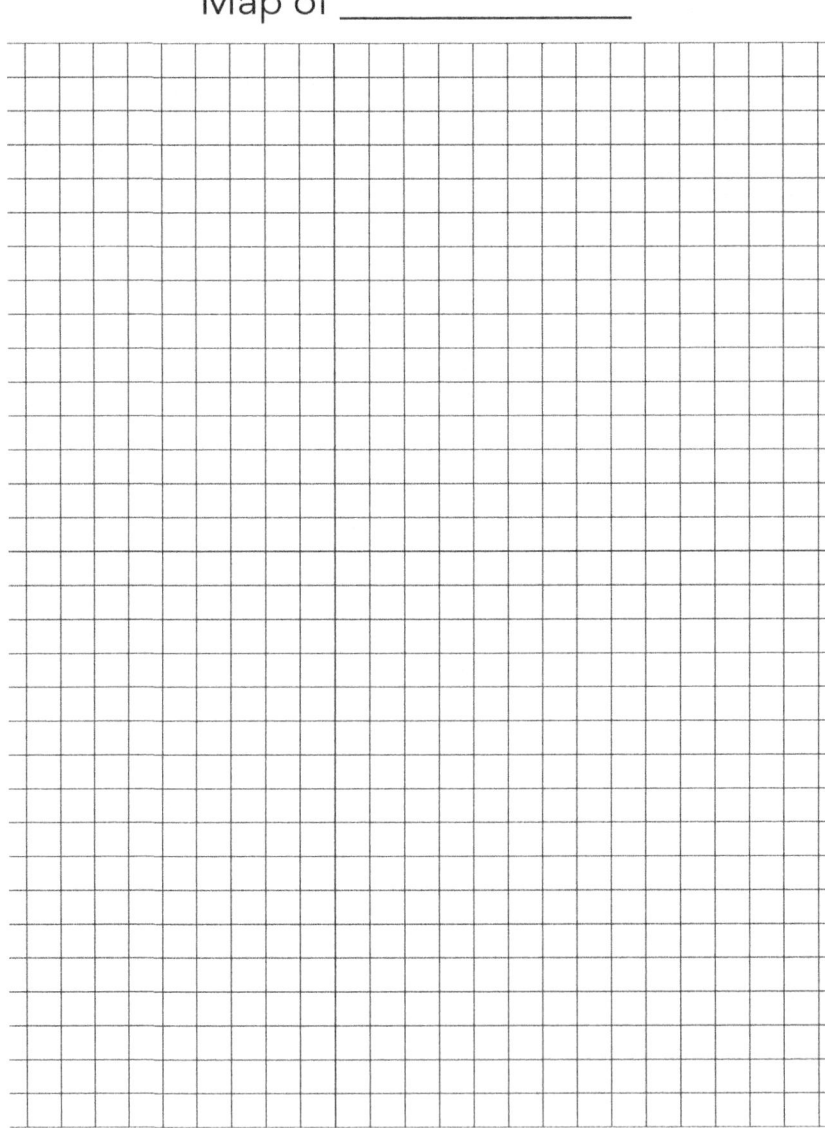

FOOD AND PROVISIONS 173

You can find these tables by signing up for my email newsletter at LauraJevtich.com/OYSV2 and you can print these out over and over.

Ideas for the Future

I decided to write this book because many of my friends were downsizing, and I wanted to help make their lives easy, stress-free, and more affordable. While minimalism does not work for everyone, it is helpful for those who want to own their stuff, not have their possessions own them. When you take emotion out of the equation and realize your stuff is just that—stuff—it is easier to purge and not buy as often.

The First Principle: Keep (This Book)

Please keep this book. But once it is no longer helpful to you, when you have gone through and done all the principles without thinking about them, give the book to someone who needs it. Or share the book with your local library. Make sure to take out the worksheets or transfer the information into the workbook.

Tip A: Worksheets and PDF Workbook

If you'd rather use a workbook, sign up for my e-newsletter and I will send it to you: Organize Your Stuff Workbook, Simple Forms to Find All of Your Stuff. This 38-page workbook has examples, worksheets, and graph paper for your most important stuff. You can go to LauraJevtich.com/OYSV2 to sign up.

IDEAS FOR THE FUTURE

Tip B: YouTube Videos

Please subscribe to YouTube.com/@LauraJevtich for videos showing how to fill in the worksheets and maps of your home so that you can find everything and everything has a place. These videos will come in the Fall of 2026.

The Second Principle: Recycle - Living Small, Downsizing, or Simple Living

I own and use my stuff wisely and never feel overwhelmed by too much. I hope this book can help you feel the same way. Use the mantra: *less is more*. By living small, downsizing to a small home, or engaging in simple living practices, you decide to have less clutter in your home. It might be a radical change or a slow, deliberate movement toward simple living. Once there, you'll no longer feel overwhelmed; you'll have more time and money, and your sanity will be saved.

The Third Principle: Sell - Knowledge Continues

Tip A: The Series - Organize Your Life

This series will comprise three books. Please stay tuned for these books to be published in 2026 by signing up for the newsletter.

Organize Your Stuff: Sustainable Ways to Declutter Your Space

All that stuff you decide to keep, recycle, sell, and donate or dump. What if your book explained how to organize and then find everything?

Organize Your Digital Files: Having a 30-Terabyte Life

Yes, you digitized everything. Now, do you know which folder or drive it is located in? This book will continue to help you find everything you digitized and stored in the proper places. After creating these systems, you will discover how easy it is to find the files you're looking for. Plus, a bonus section for those who want their own cloud system. Published June 30, 2026.

Organize Your Money: Less Means More Freedom

Each year, I encounter more and more individuals who squander funds because of a lack of financial awareness. The same principles used in the two former books apply as well to money and how you can Keep (more of your money), Recycle (ideas of old and new), Sell (to get more money), Donate (because it makes you feel good), or Dump (old ideas of work and money). It can be a taboo subject for most families and couples. This difficult subject can gain rewards with a system. You will learn about my journey with money to help you organize your thoughts, ideas, and systems. (Coming Early 2027).

Tip B: Online Courses to Organize Your Life

Online courses will start in late 2026 or early 2027 after the published series. Stay tuned by signing up for my newsletter on LauraJevtich.com.

The Fourth Principle: Donate or Dump by Recycling

When I worked as an environmental geologist, the company I worked with had a significant job in Staten Island at Fresh Kills Landfill. I worked there in the early 1990s. They wanted to know how much landfill waste leached into the groundwater.

When drilling the landfill, you had to wear steel-toe and steel-shank boots and thick gloves, as you did not know how much of the diapers, food waste, medical waste, and metals had decomposed. The clothes were the worst, as they would tangle the steel drill bits and mud together, making cleanup strenuous.

When writing this book, I returned to the subject to find the reports of what we did. I could not find a single report. The landfill closed on March 22, 2001. They started several recycling programs, including a paper recycling program, where the newspaper you read on Monday becomes a pizza box on Friday. They also capture methane gas and pump it to be recycled and sold as a profitable business for Staten Island. There is a Landfill Leachate Treatment facility on site, and what used to be a landfill is now a city park, the largest in New York City.

What does New York City do with its garbage? Staten Island has a waste facility that takes the trash, compacts it, and puts it onto trains headed for the Carolinas. This produces less waste, as organic material goes to compost facilities, the paper goes to the recycling program, and their recycling system is quite robust. They have not banned plastic bags, but several other states have, and it is only a matter of time. Many grocery stores take back plastic bags.

Tip A: Recycling is a Bad Word

Often, people do not want to take the extra steps to recycle. They would rather throw it out because it is easier to do. Out of sight, out of mind—but no. Those materials go to other states or countries. A few grocery stores and big-box stores do not offer plastic bags. Going to those stores might get other companies to change. However, banning may not be the solution. Personal change is. Realize where your garbage is going. The effort toward recycling yields worthwhile outcomes.

Tip B: Rethink Buying Stuff

What about stuff that comes in a glass container? We know that while it can be recycled repeatedly, glass still costs money and energy. What if we said no to glass? Imagine refusing Styrofoam containers entirely. Styrofoam never decomposes. Ask for aluminium foil to wrap your leftovers.

I went to a grocery store recently. I asked that the food not be put into bags, saying I had bags in my car. The cashier looked at me funny and then did what I asked. No one questioned me as I walked out of the store with the receipt in hand. Why not do that? Costco and Sam's Club do something similar, so why not have grocery stores also do it? The same could happen in regular stores, too. Why have plastic bags at all? Why not have reusable grocery bags, sewn from the recycled materials in the clothing section instead?

Tip C: Reduce the Quantities You Buy

All the principles in this book come down to a reduction in purchases. I know many crafters who shop for supplies in their own stockpiles. They did a "No Spend" month, quarter, or year. I know several who managed not to spend at all. Using up your supplies is just one way to reduce what you have. Use this book to decide what you want to keep and what you want to sell or donate.

It is a goal worth having.

This is the end of Organize Your Stuff, Sustainable Ways to Declutter Your Space, but the series is just beginning! You digitalized everything. Do you know where it is on your computer, laptop, or tablet? Or did you put it on one of those thumb drives you have in the front of your keyboard? Click here to download Organize Your Digital Files, Having a 30-Terabyte Life.

Organize Your Digital Files
Having a 30-Terabyte Life
LAURA JEVTICH

Loving the sustainable ways to organize? I would love for you to leave a review for Organize Your Stuff on Amazon or wherever you bought your book. You can also hang out with me and other book lovers in my active community. Join the sustainable, frugal, and sometimes organized community on Facebook—Join Our Community—Go to Money Mantra of Financially Frugal people.

This has the potential to be so much more than a set of fun books to read. We also have a community of creative thinkers and readers and we would love for you to be a part of it. Want to get involved? Here's how you can hang out personally with Laura and other creative minded folks. Visit Laura's community at Planners, ScrapBooks, and Journals on Facebook.

Join Laura's timely eNewsletter and receive a 38-page workbook to use with your copy of Organize Your Stuff. This twice-a-month newsletter has important news about writing, planning, and gardening. You will be the first to hear about speaking engagements, events, book news, new planner drops, organizing and decluttering household and landscape hints. Plus, what to plant and when to do so.

You can also check out my website **LauraJevtich.com** to connect with me on:

- **Facebook.com/Laura.Jevtich,**
- **Instagram.com/LauraJevtich, and**
- **YouTube.com/@LauraJevtich.**

Resources

Here are a few resources I used to create this book:

Laura and Sasha Jevtich, our lives explained in *Unusual Adventures Traveling Through Life, A Memoir of Frugalness to Financial Freedom* by Laura Jevtich, copyright 2023. This link is where you can buy the paperback: https://amzn.to/3RT44DC

The Carbon Almanac, It's Not Too Late, by The Carbon Almanac Network and Seth Godin, editor and forward, copyright 2022. You can buy the paperback or ebook: https://amzn.to/48jVGTr

Your Money or Your Life, Transforming Your Relationship with Money and Achieving Financial Independence, Joe Domínguez and Vicki Robin, copyright 1993. Paperback link: https://amzn.to/3RyIC54

Your Money or Your Life, 9 Steps to Transforming Your Relationship with Money and Achieving Financial Independence, Fully Revised and Updated for 2018, by Vicki Robin, Joe Dominguez, and Forward by Mr. Money Mustache, published 2008, updated in Kindle, 2018. https://amzn.to/470Udje

The Complete Tightwad Gazette: Promoting Thrift as a Viable Alternative Lifestyle, by Amy Dacyczyn, copyright 1998. https://amzn.to/3tgUukp

The Right to Repair, legislation, act, and bill information can be found on this website: https://www.repair.org.

Monday to Friday Cookbook, by Michele Urvater, copyright 1991 by Workman Publishing Co. https://amzn.to/3SiV4qK

Fifty Ways to Cook Most Everything, 2500 Creative Solutions to the Daily Dilemma of What to Cook, by Andrew Schloss with Ken Bookman, copyright 1992 by Simon & Schuster. https://amzn.to/43AEUye

Acknowledgements

I want to thank several people who helped make this book possible. A writing group helped: Dream Weavers Writing Club and a mastermind group: Mooseminds!

Stephanie Shackelford, Tone Eisenbise, Sharon Opperman, and Rachel Eichen were integral in suggestions and helpful hints. Both Sasha Jevtich and Robert Jevtich were invaluable in more than a few ways.

These folks helped me when I did my first version of Organize Your Stuff.

The second version, I had many beta readers and an editor as well. Plus more mastermind groups helped as well, especially when I thought I would not finish. P & T Masterminds, Romanics Mastermind, and Mooseminds, the collective group of people who came together, helped me push forward and publish. They boosted and cheered me on.

Priscilla Savary	Joan Meis Wilson
Christina Mayer	Peter Moses
Ash Reed	Esther Williams
Brenda Friedrich	Jerry Schultz
Valerie Hughes	Virtue Meadow
Tone Eisenbise	

I owe the deepest gratitude to Carla Capuano, my Mom, who helped edit the first draft in 2018. She and my Dad, Louis Capuano, suggested I publish my memoir first and tackle the series later.

Thank you all, with hugs and happiness always.

About the author

Laura Jevtich works as a writer, author, and social media creator. She enjoys being part of a community, whether at home, online, or traveling. She is a Master Gardener teaching others to chase the shade, and she strives to read, write, and play well with others. You can find her in Southern Nevada, with her husband, Sasha, and online: @LauraJevtich for author information and creative productivity.

To contact her: Laura@LauraJevtich.com or LauraJevtich.com

Love hanging out in my community! It is one of the best parts of being a writer, and I'd love for you to be a part of either my frugal sustainable community at **Money Mantra of Financially Frugal** Facebook group or for creatives at Planner, ScrapBook, and Journals Facebook group.

Want to know more about me, what Sasha (my husband) and I do to have a sustainable lifestyle, how I developed my writing process, or the plants we try to grow in the desert? Sign up for my eNewsletter and I'll tell you all my frugal secrets! I'll even send you a workbook you can use with this book. In the meantime, keep reading! Grab my memoir to see how I developed into an organized, sustainable, and environmentally conscious person.

Connect with Laura Online:

FacebookFacebook.com/laura.jevtich
InstagramInstagram.com/laurajevtich
ThreadsThreads.com/@laurajevtich
YouTubeYoutube.com/@LauraJevtich
BookbubBookBub.com/profile/Laura-Jevtich
GoodreadsGoodreads.com/laurajevtich
Website LauraJevtich.com

Free Workbook for Organize Your Stuff book and ebook!

ABOUT THE AUTHOR

There are 38 pages to help you write where you placed your stuff. Available only to eNewsletter subscribers, sign up to get it: at: https://laurajevtich.com/#newsletter

Also by

If you loved this book, you will definitely love her other books and journals. Easy-to-understand and uncomplicated steps to use her methods.

Unusual Adventures Traveling Through Life

How she went from a 3000 square foot house to a 24 square foot RV with her husband. The best how-to for living small, living your passions, and staying with your spouse throughout it all.

Available on Amazon

ALSO BY

Gardener's Journal of USDA Hardiness Zones 8 & 9:

Tips and tricks for gardening in the desert every month of the year. Dated for 2026, with moon cycles and celestial events throughout the year.

Available on https://laurajevtich.gumroad.com

PLOW'G Method & Mission Statement Creation

How does Laura do it all and take a snooze or create in her junk journal or planner? Download this free workbook to find out. Using your own tasks for the four aspects of your life, you may find you have more time to do what you want and create a personal mission statement along the way.

Available on https://laurajevtich.gumroad.com

Please sign up for my newsletter at LauraJevtich.com to get news about the series, gardening, planning, and writing tips.

Made in the USA
Coppell, TX
27 February 2026

72448671R00115